ELITE SOULS

ELITE SOULS

PORTRAITS OF VALOR IN IRAQ AND AFGHANISTAN

RAYMOND JAMES RAYMOND

NAVAL INSTITUTE PRESS
ANNAPOLIS, MARYLAND

Naval Institute Press
291 Wood Road
Annapolis, MD 21402

ISBN 978-1-68247-713-7 (hardcover)
ISBN 978-1-68247-788-5 (eBook)

Library of Congress Cataloging-in-Publication Data is available.

♾ Print editions meet the requirements of ANSI/NISO z39.48-1992 (Permanence
of Paper).

Printed in the United States of America.

30 29 28 27 26 25 24 23 22   9 8 7 6 5 4 3 2 1
First printing

To my beloved Kathy, without whose
loving support this book would not
have been written.

CONTENTS

PART 3. PREPARATION

INFANTRY

AVIATION

PART 4. INTO BATTLE

PART 5. EPILOGUE

ACKNOWLEDGMENTS

In writing this book I have accumulated many debts. The first is to the five elite souls: Maj. Nick Eslinger, Maj. Tony Fuscellaro, Maj. Ross Pixler, Maj. Bobby Sickler, and Maj. Stephen Tangen. By agreeing to cooperate with me on the book, they trusted me to tell their story accurately, objectively, respectfully, and with empathy. I hope I have done so and proven worthy of their trust.

I also owe an immense debt to my old friend, Col. Bob McClure, USA (Ret.), former president and CEO of West Point's Association of Graduates, which bestows the Nininger Medal annually. Bob very kindly agreed to support my project and introduce me and it to Nick, Tony, Ross, Bobby, and Stephen, all Nininger Medal recipients. Without Bob and his generous support, I could never have begun the project.

That is equally true of my friend Col. Suzanne Nielsen, chair of the Social Science Department at West Point. Suzanne's role was vital. She not only advised and encouraged me but also wrote a generous letter of recommendation for me to Nick, Tony, Ross, Bobby, and Stephen. I am forever in her debt. Not content with that, Suzanne recommended me and my project to Joe Craig, director of the Book Program at the Association of the United States Army (AUSA), when he visited the Social Science Department at West Point. In turn, Joe kindly took on a first-time author and helped me revise my book proposal for the Naval Institute Press, with which he had developed a new partnership for AUSA authors. At the press, Glenn Griffith, acquisitions editor, showed similar kindness in offering me a publication contract. He has also responded to my numerous queries with patience. I owe a special debt to Yvonne Ramsey for her meticulous copyediting and for her patience with an author who struggles mightily with technology.

A number of good friends helped me define the project: Col. Jack Jacobs, USA (Ret.); Brig. Gen. Cindy Jebb, USA; Brig. Gen. Dan Kaufman, USA (Ret.); Col. Jay Parker, USA (Ret.); and Professor Doug Stuart. I am forever in debt to the friends who patiently read and reread parts of the manuscript and offered the kind of frank, unvarnished, and incisive comments I needed: Dr. Darryl Banks; Dr. Sky Foerster; Brig. Gen. Jim Golden, USA (Ret.); Brig. Gen. Dan Kaufman, USA (Ret.); Col. Jay Parker (Ret.); and Judge Al Rosenblatt, New York State Court of Appeals (Ret.). Five other friends offered constant encouragement and support: David Buckley, Khalil Habib, Mike Giglio, Dan Hottel, and Jonathan Weinstein.

I must also thank Jim Ludes, director of the Pell Center at Salve Regina University, for kindly arranging a Pell Honors Forum in which I could test some of the book's main ideas with an invited audience.

At West Point I must thank Vice Dean John Dalton, who arranged access to relevant documents in the United States Military Academy's archives, and Meghan Dowers-Rogers, who scanned all the documents and made them available to me electronically.

Thanks are also due to Max Cottet, my technical assistant, who helped me prepare the manuscript for the publisher.

The biggest debt I owe is to my wife, Kathy. Throughout all of the time I have been working on the book, she has given me not only loving support but also sound advice and wise guidance. I owe her a debt I can never repay. Our two sons, William and Nathaniel, have also been stalwart supporters of their dad's book.

A NOTE ON METHODOLOGY

Most of this book is based on multiple extended interviews with each of the five elite souls conducted between 2016 and 2020. I also interviewed family members, friends, coaches, mentors, and, where possible, superior officers under whom the five officers served, as well as non-commissioned officers and soldiers who served under them. From the beginning of the project, I pledged to the five officers an account of their lives, education, training, and battlefield experiences that would be meticulously accurate, frank, and unvarnished. To that I end, I invited the five officers to fact-check the manuscript. The final judgments are mine. Nothing has been left out. That was my goal and theirs.

INTRODUCTION

Absolute bravery, which does not refuse combat even on unequal terms, trusting only in God or destiny, is not natural in man. It is the result of moral culture, and it is infinitely rare.... Among the elite souls, a great sense of duty that only they can understand and can obey is supreme.

—*General Ardant du Picq,* Battle Studies, *2017 (1868)*

War is not noble. Sometimes it may be necessary and even justified. But it is always a savage struggle for survival or supremacy in which young servicemen and now servicewomen see their comrades, in the words of John Ellis, "shot through the eye, ear, testicles, or brain" or have their legs blown off by a mine, an exploding shell, or an improvised explosive device. As he also wrote, "combat is torture ... which will reduce you sooner or later to a quivering wreck."[1]

In war there is always danger, death and destruction, fatigue, fear and frustration, hardship and horror. Very often, there is also comradeship and courage, excitement and exhilaration. Above all, there is physical and psychological pain. The physical wounds usually heal faster

1

than the psychological: the intense vivid flashbacks to combat that linger so long, the memories of comrades killed or wounded, the faces of the enemy you killed, the homes and businesses you had to destroy. Most painful of all are the memories of the innocent who died, caught in a cross fire or as a result of a bomb that went astray.

Wars take a terrible toll on everyone who has to fight them. To one degree or another, war robs servicemen and servicewomen of a part of their humanity. As a result, we owe them all a profound debt. As author Mark Aronson correctly suggests, "We owe them the respect of listening to them. We have to honor their experience by paying attention to it, no matter how uncomfortable it makes us. They have the right to curse at us, at life, at fate. They have the right to be bitter; they have the right to nurse their wounds."[2]

Successfully meeting these dreadful realities of combat demands a special brand of courage. In the famous words of Gen. Omar Bradley, "Bravery is the capacity to perform properly even when scared half to death." General Bradley was right, but only up to a point. The British Army's General Sir Peter de la Billière put it better when he wrote that "moral courage is higher and rarer in quality than physical courage. It embraces all courage, and physical courage flows from it."[3] In other words, courage is not synonymous with bravery. Criminals and terrorists can be brave. True courage is more complex. As West Point English professor Elizabeth Samet suggests, "Choosing the harder right requires a quality beyond physical bravery."[4] In other words, in war true courage is selfless and serves a higher moral purpose. An act of true courage in combat can mean saving the lives of your buddies or those under your command. It can also mean making an important contribution toward winning an engagement with the enemy or denying them a key tactical or strategic objective.

Author Sebastian Junger has also written eloquently about the true nature of courage. In his classic *War*, he writes that "among other things, heroism is a negation of the self—you're prepared to lose your own life for the sake of others."[5] He is right.

I believe that Gen. Ardant du Picq, the distinguished nineteenth-century French military thinker, is spot-on when he argues that this kind of selfless courage is rooted in "moral culture" and is to be found among "the elite souls."[6]

This is a book about five elite souls—Nick Eslinger, Tony Fuscellaro, Ross Pixler, Bobby Sickler, and Stephen Tangen—all highly decorated young West Point graduates who personify the kind of selfless moral and physical courage described by Ardant du Picq.

This book not only tells the life stories of these five elite souls but also examines what Lord Moran, Winston Churchill's wartime physician, once called the "anatomy of courage" that motivated their selfless actions. *Elite Souls* does not pretend to offer new concepts. I have not developed a new theoretical concept that has eluded every student of the subject from Thucydides to the present day.

What I have done is take a theoretical concept set out by distinguished military thinkers—that physical courage flows from moral courage and strong moral values—and applied it. The book is therefore a case study of how this concept was developed in five exceptional officers and has shaped their lives and military careers.

Conceptually, the book is also a case study of concepts of moral leadership and character building developed by *New York Times* columnist David Brooks in a series of classic books including *The Road to Character* and *The Second Mountain: The Quest for a Moral Life*. He has developed them in a civilian context. I have applied them to the military.

The central argument of this book is that the five elite souls each embody a marriage of outstanding ability with exceptional moral character. Each of them entered West Point as young men of great promise with impressive records of intellectual, leadership, and physical achievements. They already had an unusually strong moral foundation thanks to remarkable parents, outstanding mentors, and a strict but loving upbringing. West Point's rigorous education, military training, and leadership development model built on that foundation to produce young military leaders who personified the classical virtues: moral and physical courage, self-sacrifice, a deep commitment to duty, personal and professional honor, and selfless service to the nation. They are among the very best that West Point produces.

This is not to suggest that West Point was perfect. Far from it. Like any human institution, West Point was (and is) imperfect because it is led by and attended by people who are themselves imperfect. West Point has always had challenges. Because it admits cadets from every part of the United States, it inevitably imports in each class the tensions and fissures that exist in American society. It has always been West Point's job to ensure that whatever values entering cadets have, they are replaced with a set of values found in leaders of character, integrity, honor, and professionalism. When the five elite souls were cadets, there were three challenges facing West Point: friction over the presence of women in the corps, sexual harassment and assault of female cadets, and alcohol abuse. The United States Military Academy (USMA) made progress in dealing with these challenges, but when these elite souls graduated, much work remained to be done. Looking back, these men saw their West Point experience as transformative but imperfect. It was a slow, grinding process that demanded all of their determination and resilience to overcome the inevitable frictions and setbacks. In the end, West Point succeeded in transforming five remarkable young men into principled leaders of character well equipped to deal with the moral complexities and military challenges of the wars in Iraq and Afghanistan.

In the field army, their noncommissioned officers (NCOs) advanced their character-building process. Among the NCOs, Pete Black, Ross Pixler's platoon sergeant, put it best when he said at their first meeting, "The men come first no matter what. Never compromise the respect and integrity of the platoon. You don't take from my men." Ross never did, and neither did any of the other elite souls. They were challenged, inspired, and trusted by their superior officers, leaders who were emphatic about clear ethical standards and the need to fight according to the laws of war. The battalion or squadron commanders were inspirational leaders who exuded selflessness and also helped inspire these young lieutenants to achieve the highest levels of professionalism and perform acts of uncommon valor on the battlefield.

Let's meet each of the elite souls. First, Lt. Nick Eslinger, a tall, brown-haired twenty-four-year-old from the San Francisco Bay area. Nick was a 2007 West Point graduate and a 2008 graduate of the U.S. Army's rigorous Ranger School.

On the evening of October 1, 2008, Nick was on patrol with his platoon in a narrow walled lane in Samarra, Iraq. An insurgent threw a hand grenade over the wall. Nick dived on the grenade to protect his soldiers. Fortunately, it did not explode when it was supposed to, and Nick had time to throw it around the corner of the alley wall. This was his first contact with the enemy. He had only been a platoon leader for a few months and had only been in Iraq since July. Nick received the Silver Star in recognition of his selfless courage, later upgraded to the Distinguished Service Cross, the nation's second-highest award for valor.

Second, meet Lt. Tony Fuscellaro, a tall, black-haired twenty-six-year-old from Fearless, Pennsylvania. Tony was a 2005 graduate of West Point. By August 2008, he was a Kiowa pilot in command and had been serving in Kandahar Province in Afghanistan for four months. Kiowas are small two-seater scout armed reconnaissance and close air support helicopters. Their role is to stay low and fly fast in support of ground troops.

On August 24 eight miles from Kandahar City, Tony flew his lightly armored Kiowa into fierce enemy fire to save the lives of twenty Army engineers ambushed by over one hundred Taliban. Out of ammunition and low on fuel, Tony flew at fifty feet into a hail of small-arms and antiaircraft fire while firing his personal carbine to distract the Taliban, thereby enabling the engineers to escape. Tony received a Distinguished Flying Cross in recognition of his selfless courage. Later in the same deployment Tony received a second Distinguished Flying Cross and two Bronze Stars for further acts of selfless valor.

Third, meet Lt. Ross Pixler, a tall, rugged mountaineer from Phoenix, Arizona. Ross was a 2005 West Point graduate and the son of an exceptionally courageous U.S. attorney who had prosecuted the Mexican drug cartels and survived. A 2006 alumnus of Ranger School, this was Lieutenant Pixler's first deployment to Iraq, part of President George W. Bush's new surge strategy.

On October 30, 2007, near Al Bawi, Iraq, insurgents exploded a massive improvised explosive device (IED) under Ross's Bradley Fighting Vehicle, killing three of his soldiers and seriously wounding Ross, his driver, and his gunner. Suffering from shock, a concussion, a brain injury, and a seriously damaged neck, Ross saved his wounded soldiers' lives. He then led his patrol through an intense three-hour firefight with the insurgents until reinforcements arrived and the enemy withdrew. Ross received the Silver Star in recognition of his selfless courage.

Fourth, meet Lt. Bobby Sickler, a humble, quiet, twenty-seven-year-old from rural West Virginia. Bobby is the son of a distinguished U.S. Marine Corps officer and heir to a family tradition of military service dating back to the American Civil War. Bobby graduated from West Point in 2005 and deployed to Iraq in June 2007. By December 2007, he was a battle-hardened veteran of forty firefights with insurgents on the ground. This suited him just fine. Behind his quiet, almost priestly demeanor was a fierce warrior who, after graduating from flight school, had wanted to join whichever unit was going to war next so, he said, he could "mix it up with the enemy."

For six hours on December 30, 2007, Bobby tracked and eventually killed a group of Iraqi insurgents trying to escape with a vehicle inside of which was a ZSU heavy weapon that could have destroyed his squadron's unprotected Kiowa helicopters. While Bobby was flying at less than one hundred feet over Mosul, his Kiowa was hit by heavy ground fire. Despite his failing engine and flight instruments, he successfully flew his badly damaged Kiowa back to base. There he landed and without orders took off again in his squadron's spare Kiowa to complete his dangerous mission. Bobby received the Distinguished Flying Cross for his courageous action.

Finally, meet Lt. Stephen Tangen. A muscular, brown-haired twenty-four-year-old champion swimmer, Stephen grew up in Naperville, Illinois, a prosperous western suburb of Chicago. He was a 2008 West Point graduate who had taken command of his rifle platoon from

Nick Eslinger just a few months earlier. Before late June 2009, Stephen had not experienced either direct enemy fire or any type of combat situation with the enemy.

On June 27, 2009, Stephen's platoon was the spearhead of a joint U.S.-Afghan operation to clear the Ghaki Valley of Taliban insurgents who had flooded into eastern Afghanistan. Despite the tightest security, Stephen's platoon was ambushed on a narrow one-lane road at a point where they could not turn their vehicles around and there was no cover; thus, the Taliban could zero in on every weapon they had.

For the next twelve hours, Stephen gave an extraordinary example of leadership and selfless courage in an intense firefight in temperatures exceeding 100 degrees. As he organized the defense of his dangerous, exposed position, he put new life and resolution into his grief-stricken young soldiers, only four of whom had prior combat experience. On what must have seemed like the road to hell, Stephen was indefatigable. He received the Silver Star in recognition of his selfless courage.

Each of these five elite souls are shining examples of selfless service to the nation. Each of them are highly decorated combat veterans and recipients of West Point's Nininger Medal: Nick in 2009, Bobby in 2010, Ross in 2011, Stephen in 2012, and Tony in 2013.

The Nininger Medal is named in honor of 2nd Lt. Alexander Nininger (West Point class of 1941), who received the first Medal of Honor of World War II. Nininger died fighting the Japanese in the Battle of Bataan in July 1942. The medal was endowed by E. Douglas Kenna (West Point class of 1945) and his wife, Jean Kenna, and is awarded annually by West Point's Association of Graduates.

In a way, the Nininger Medal is West Point's own medal of honor. The award recognizes exceptional courage and is given to "an exemplar of heroic action in battle." In addition to recognizing recipients for bravery as individuals, the Association of Graduates regards the recipient "as a given year's representative of all the West Point–commissioned officers who have heroically led soldiers in combat." A joint West Point–Association of Graduates task force reviews the military records of all graduates who have been decorated for bravery in a given year and chooses the annual Nininger recipient. The task force normally chooses a relatively recent graduate currently on active duty who best represents West Point's noble values of duty, honor, and country.

Recipients of the USMA's Nininger Medal have known the cold, gnawing fear of death in combat; the pain of long separation from family; and the death of comrades. Our five elite souls never set out to become heroes or win glory for themselves. Rather, they wanted to be first-class professionals with a sound moral compass. Their combat experiences were frightening and intense. Yet somehow despite the trauma and horror of war, they found the courage to do way more than their duty, putting their lives at risk to save others.

These five elite souls, all Nininger Medal recipients, also have a wider significance. They embody the most selfless and heroic qualities of the millennial generation who were in elementary, middle, or high school on 9/11. Those attacks not only traumatized a nation but also transformed the lives and the outlook of this generation. Multiple studies suggest that 9/11 created a more purposeful, focused, and civic-minded generation for whom serving others is a higher priority than it was for preceding generations. Over 2.8 million American millennials have volunteered to serve in the U.S. armed forces; 13,000 of them were accepted to West Point. After four years of rigorous education and military training, most of those graduates led troops in combat in Iraq and Afghanistan. The Nininger Medal recipients represent the most courageous and selfless of this new generation of American war fighters who graduated from West Point.

I hope that civilian readers of this book will gain a deeper understanding of the complex human realities of modern war. As West Point professor Elizabeth Samet has written, "Television, embedded reporting, and videography have turned the rest of us into war's insulated voyeurs. . . . While technology creates an illusion of intimacy, the consumption of the war as a spectacular movie arouses our pity but does nothing to enlarge our sympathies."[7] I would add that technology also does nothing to enhance our understanding. I believe that a full analysis of individual acts of exceptional courage by the remarkable young junior officers who fought their nation's wars in Afghanistan and Iraq can help the American public deepen their understanding of the human realities of modern war.

Finally, this is a tribute. The United States was at war in Afghanistan, and later Iraq, for close to twenty years until President Joe Biden ended the Afghanistan War in August 2021. These wars were fought by the bravest young men and women most Americans will never know. I know some of them; I have taught them at West Point. I especially want Americans to meet these five Nininger Medal recipients, these elite souls who embody the true spirit of courage. This book is my humble tribute to five of the most remarkable human beings I have ever met.

PART 1

EARLY LIFE

1

NICK ESLINGER

Growing up, Nick Eslinger was a quiet, reserved boy and a very intelligent and successful student-athlete. He was a blue-collar lad from a devoted hardworking blue-collar family in the San Francisco Bay area. Nick was your classic straight arrow: a bright student with a tremendous work ethic and a strong moral compass, a natural leader who became captain of his high school football team.

Nick was born in Mountain View, California, on February 21, 1984. Mountain View, nestled between the Santa Cruz Mountains and San Francisco Bay, is located in the heart of what is now Silicon Valley, ten miles north of San Jose and thirty-five miles south of San Francisco. At the time, Mountain View was a rural community with lush orchards and hundreds of acres of vineyards. In 1993, Nick and his family moved to Oakley, California. Sitting on the banks of the San Joaquin River, an hour east of San Francisco, Oakley was a beautiful place to grow up. Later, Nick's mother, Donna, said that "raising the boys in this small town was a blessing." Oakley—the name comes from Old English and means "a meadow of oak trees"—still had many beautiful old oaks when Nick and his family arrived there. But by the 1990s, Oakley was also a boom town driven by the expansion of nearby Silicon Valley.

Between 1990 and 2000, the population of Oakley increased from 18,374 to 25,619.

Nick's parents, Donna and Bruce, both had difficult lives but grew to be remarkable models of resilience, steadfastness, and selfless devotion to family. Donna was born in Lubbock, Texas. When she was thirteen, her father moved the family to Santa Fe in northern California. Her parents divorced, and her father abandoned the family, leaving Donna's mother to raise three daughters alone. Life was a constant struggle. As the eldest, Donna felt a special responsibility to help her family. Once she graduated from Willow Glen Senior High School in Santa Fe, she went straight to work to support the family. In the evenings, she took sign language classes at a local college. In 1982 she married Eric Eslinger, Nick's biological father, but they were both very young; they grew apart and divorced after three years. Nick was two years old. Four years later, Donna married Bruce Behnke.

Bruce was born in Santa Fe, California, and grew up the child of a single mother. He overcame many challenges to become a strong, caring, successful man. He began a thirty-year successful business career in California with Cardinal Health Care, a leading *Fortune* 500 company headquartered in Ohio that specializes in providing custom solutions for every branch of the U.S. health care system ranging from hospitals to doctors' offices and pharmacies.

Bruce, a steady, kind, and loving man, was devoted to not only Donna but also Nick. And although Eric moved three hours way and was not a presence in Nick's day-to-day life, he remained a dedicated and loving father who regularly drove six hours round trip to attend Nick's sporting events. By this time Donna and Bruce had their own child together, Danny. Nick and Danny bonded as brothers, and together the four formed a warm, loving family. This kind of emotional stability proved vital to Nick's development.

Nick grew up in a modest 1,600-square-foot track home on a quarter-acre lot in Oakley. Donna and Bruce were not college graduates, but both became successful professionals. Donna taught health and fitness classes at the Oakley YMCA before becoming a personal trainer. Bruce became a facilities manager with Cardinal Pharmaceuticals and later was director of compliance and safety.

Unlike the other elite souls, organized religion played no real role in Nick's family life. But good values mattered. Donna had many conversations with him about kindness, patience, and respect for other people. Overall, Nick's parents always taught him that good habits and good values breed good character. They counseled him to always do the harder right, to always do the selfless not the selfish thing.

As a boy growing up, Nick was surrounded by inspiration and excellent role models. His parents inspired him by their ethics, hard work, and selfless devotion. For ten years Donna was the main parental presence, because she worked days while Bruce worked nights, usually getting home at 4 a.m. Donna established a predictable and orderly schedule for Nick and Danny, four and a half years Nick's junior: school, sports, homework, dinner, sleep. On weekends, Donna, Bruce, and the boys went waterskiing. Eric also contributed. Most Friday nights, for example, when Nick was in high school, Eric drove from Placerville, near Lake Tahoe, to Oakley to watch Nick play football. Although their budget was limited, Donna and Bruce always found a way to pay for opportunities that would benefit their sons. As a child, Nick was surrounded by selflessness and excellent role models. In his application to West Point, he paid a moving tribute to his mother's influence in shaping his character:

> Throughout my life, especially my teenage years, many people have provided me with guidance and education. The one individual, however, who has influenced me the most, has been my mother. For as long as I can remember, she has taught me life-lessons, courage, and most of all respect. Furthermore, she has always been there to answer any questions I may have. The one lesson I remember most is to never break a promise. My mother would rarely promise me anything, but when she did, she always came through. I value her love for me more than anything in the world. I hope to have a similar influence on my children one day.[1]

Another source of inspiration was Nick's best friend, Matt Huffaker. Nick and Matt first met in seventh grade English class at O'Hara Park Middle School. They were two high-energy boys who bonded quickly. Their friendship was built on a passionate interest in sports, a similar sense of humor, and above all similar ethical and moral values. Matt was the son of a family court judge and a schoolteacher. Matt's father was a man of high moral character who lived a life of unimpeachable integrity, inspiring his son and Nick by his example. As a result, Matt always did the right thing and constantly challenged Nick to run more, study more, and practice more. Nick told me that he "naturally assimilated to Matt's behavior." This is true, but Nick also inspired Matt's actions. Together they reinforced each other's honorable ambitions, ethics, discipline, determination, resilience, and self-discipline.

Freedom High School's charismatic football coach, Larry Rodriguez, added a new level of example and inspiration. Rodriguez first met Nick in September of 1998 in his first-year PE class and was amazed. Nick ran laps with such determination and energy that he finished one hundred meters ahead of his classmates. Later Rodriguez described Nick as "one of the top athletes I ever encountered in thirty-six years of coaching." Rodriguez taught Nick and his teammates to play with honor and integrity, to win in a way consistent with the selfless values he taught them. Cheating and foul play were unacceptable. At the same time, Rodriguez encouraged and nurtured Nick's leadership skills. The coach never micromanaged and always emphasized to Nick that "it's not about you, it's about the team," meaning that as team captain Nick had to step up and make sound decisions for the good of the team.

And he did. As the starting quarterback and captain of Freedom High's football team. Nick was always the first to arrive for practice, determined to practice hard and encourage his teammates to do the same. In Nick's first season as captain, his leadership helped transform a losing team into the best in the division. He also led by example. Only a few days after being released from a neck brace following an injury, Nick insisted on playing because he would not let his teammates down. He loved his teammates like brothers. Later Coach Rodriguez told me that "Nick was a born leader. I didn't have to do much; I just added a bit to the recipe."

Like most of the other elite souls, there was no military tradition in Nick's family, and neither of Nick's parents suggested that he might join the Army. By the time he was a junior in high school, Nick had never heard of West Point. His family had no plans for his college education, although Nick wanted to go to college and had begun seriously thinking about it. As he told me later,

> I wasn't one of those kids who had my whole life planned out. I was an athlete, and sports were my priority; my near-term focus was determining whether I would be good enough to play at the collegiate level. My junior year is when I had a serious conversation with my coaches—both golf and football—and determined I was not quite there. So at that point, I had two options: Stay local, continue living with my family, and work on my athletic skills so that maybe I could get a scholarship down the road or maybe join the military.

Nick's road to West Point began unexpectedly. As he and Matt were walking toward the Freedom High parking lot one spring afternoon, they saw two cadets in their formal gray uniforms and stopped to chat with them. This is how Nick remembered the encounter:

> I had no idea who they were, but they were in their gray uniform, and for some reason I stopped and I asked them who they were and what they were doing. I don't know why, but something drew me to them. I think it was the uniform. They said, "We're here from West Point." I said, "What's West Point?" They said, "Well, if you are not doing anything at one o'clock, come to our presentation in the career center. We're here just to spread the word about West Point." I looked at Matt, and he said, "I know what West Point is. Let's go." Matt and I went to the presentation that afternoon. He and I were the only students to come, and by the end of it I wanted nothing else but to go to West Point. I was hooked. They just conveyed leadership, and

they looked like they had it all together. They looked like they were just bound for success, and I wanted that. It was my first exposure to a military academy. Contributing to my desire to go there was the fact that Matt wanted to go there too, so if I could go somewhere where my best friend was going, we could do this together. That was really the first day I ever heard of West Point, and my decision about what I was going to do after high school was made. Now it was a matter of figuring out what I needed to do to get to West Point.

Nick left that meeting with the two West Point cadets not only with a burning desire to serve his country but also with a fierce determination to earn a place at the USMA. Later, Matt said that "Nick was born for West Point. This was God connecting Nick with what he was meant to do."

Donna was stunned; she never envisaged Nick becoming a military officer. He had always loved military history and enjoyed discussing the great wars of the past with Bruce. But up until this moment, Nick had shown only a limited interest in military service and never considered West Point. But once he convinced his parents that this was really what he wanted to do, they fully supported his choice. Nick and his mother became a team, spending hours together researching every aspect of the demanding West Point application process so he could succeed. As Donna recalled, there was "a lot of sweat and tears getting him to West Point." She was a loving but firm taskmaster, constantly reminding Nick that "if you want to live your dreams, this is what you have to do."

Like the other elite souls, moral reasons were important in driving Nick toward West Point and serving as a military leader. As he wrote in his USMA application, for him it was a way to live "a life of meaning and purpose" that would "make the world better for having lived in it."[2]

The USMA loves scholar-athletes with good values and real leadership experience, so Nick looked like a strong candidate. In his teenage years, he had begun to excel as a student and as an athlete. In seventh grade at O'Hara Park Middle School, an inspirational teacher, Mrs. Benedetti, awakened in Nick a love of learning. In high school, Nick excelled in advanced placement (AP) classes. In AP calculus, arguably the hardest class in Freedom High School, his teacher Kevin Allen commented, "I appreciate that Nick gives me one hundred percent of himself every day. He always strives to understand the concepts. He doesn't take days off. He has strong goals and he expects himself to achieve them."[3] Nick's English teacher, Mark Gates, described Nick as "an insightful writer who has contributed a great deal to his AP class."[4]

By then, it was also clear that Nick was an unusually gifted athlete who broke Freedom High School records and captained the golf team and, of course, the football team. In short, Nick had everything West Point sought in a cadet: the grades, the values, and the leadership experience. Jeff Jonas, a teacher and coach at Freedom High, described Nick as "one of the finest student athletes I've had the pleasure to work with."[5] Nick's one academic weakness was that he struggled with the SAT, and his initial score was just below West Point's requirements.

The key to success for admission to West Point lay in getting a nomination by a member of Congress. Nick completed the paperwork and got an interview with Congresswoman Ellen Tauscher's staff. A centrist Democrat on the House Armed Services committee, Tauscher was a strong supporter of the U.S. military. Nick got a nomination, number two out of Tauscher's allotment of ten. This meant that he had a good chance of getting into West Point, but success was not guaranteed. About six months later, Nick's family received a telephone call from West Point's admissions office telling them that Nick was a strong candidate, but as of yet West Point could not offer him a place. Instead, the admissions office offered him a place at the United States Military Academy Preparatory School (USMAPS). Nick and his parents had twenty-four hours to make a difficult decision. To accept the prep school placement would be to abandon his attempt at direct admission for that year. To decline it would be to risk losing the opportunity to go to West Point. Nick and his parents decided not to risk anything. The USMAPS was a sure thing, and they took it. There, Nick would raise his SAT scores and learn how to be a cadet.

In the summer of 2002, Nick arrived at the USMAPS. Founded in 1946, by 2002 it was located in Fort Monmouth, New Jersey, in a two-story redbrick Bauhaus-style building that had once housed the U.S. Army's Signal School. Then and now the school's mission was to prepare its students academically, militarily, and physically for West Point. The students included enlisted soldiers nominated by their company commanders and, since 1965, outstanding young high school athletes and leaders nominated by their congressmen and senators. For everyone, the first month at prep school is demanding. It consists of three weeks of basic training to introduce cadet candidates to basic soldiering skills and to develop mental and physical resilience. The final week prepares cadets for the academic year ahead.

At first the USMAPS was a bit of a shock, but Nick loved it. He excelled in the first three weeks during boot camp. For him, it was fun to challenge himself physically and mentally. He was physically fit and thrived, but the mental challenge was something he had never yet encountered. Staff Sergeant Orloff and Capt. Mark Manns, two outstanding tactical advisers, helped him meet that challenge successfully. Their inspirational leadership, integrity, and selflessness left an indelible imprint on Nick. "I want to be just like them," he decided. And he turned out to be. In support of Nick's application to transfer from the USMAPS to West Point, Captain Manns wrote that Nick is "a dedicated and conscientious young man. He is a capable and charismatic leader who possesses the ability to succeed in all endeavors. He is a leader and role model among his peers in the company. He is an outstanding soldier, an intelligent and hard-working student, and an exceptional athlete who excels at USMAPS. Without a doubt he has what it takes to succeed at West Point."[6]

As a bright student, Nick found the USMAPS academic curriculum easy to master, and with more intensive preparation he raised his SAT scores. What was really important to him was the opportunity to develop into an effective military officer. As Nick remembered,

The number one thing the prep school does for its cadet candidates is to teach them how to be a cadet and how to be in the Army. The benefits of the prep school were operationalized in my squad during Cadet Basic Training. I was the only prep school graduate in my squad, so when it came time to complete simple tasks like tying your gear down or assembling your rifle, I was able to share the tacit knowledge required to perform these tasks appropriately, and they started looking to me for just basic guidance. And so naturally through having that tacit knowledge, it allowed me to become an informal leader among our squad members. That fact put me in positions of leadership at an early stage and allowed me to learn and practice leadership. It's all part of the system of developing leaders at West Point.

The USMAPS set Nick up for success as a cadet, a soldier, and a leader.

2

TONY FUSCELLARO

At age eighteen, Tony Fuscellaro was a handsome young lad, the son of devoted Italian American parents. Five foot nine with a friendly face, a warm impish smile, black hair, and striking brown eyes, Tony was from a hardworking blue-collar family. He was also a deeply religious young man from a devout Catholic family with an unusually strong moral character deeply committed to serving others. Tony was a straight arrow: a studious, self- assured, intellectually gifted boy with a fierce work ethic determined to succeed, with no run-ins with school or other authorities. He was dedicated, organized, persistent, personable, and trusted by everyone.

Born in Fairless Hills, Bucks County, Pennsylvania, in 1983, Tony grew up in this proud steel town, a forty-five–minute drive from Philadelphia and ten minutes from Trenton, New Jersey. Tony's parents, Gina and Anthony, came from humble beginnings, but what they lacked in wealth they more than made up for in moral character. Anthony and Gina were a deeply religious couple, people of high moral principle and unimpeachable integrity. They were patriots: deeply devoted to the well-being of their family, their community, and their nation.

Gina was born in Bensalem, Pennsylvania, and grew up in Nottingham, one of its neighborhoods. She and her family were devout

Catholics, members of the Our Lady of Fatima Parish where she attended public schools. After graduating from high school, Gina worked in the catering industry.

Tony's father, Anthony, was born in Huntingdon Valley, Pennsylvania, an outer suburb of Philadelphia, and also grew up in nearby Bensalem. Anthony attended our Lady of Fatima School through eighth grade before graduating from Bensalem High School. After graduation, he became a mechanic and later an industrial engineer. He studied at the Claver Buckner School and Rutgers University, in New Jersey, where he qualified as an operational engineer with professional licenses in industrial engineering and wastewater management. He joined the State of New Jersey's engineering force in 1984.

Gina met Anthony, her future husband, when they both acted, danced, and performed in amateur variety shows with the Fatima Follies to benefit Our Lady of Fatima Parish. After they married in 1982, Gina became a homemaker, but when Tony and his younger sister Alicia Marie (b. 1985) went to school, Gina returned to work part time. Every evening she and her manager, constantly bent over, dry-cleaned carpets in department stores and office buildings. Worse, on a family trip to Busch Gardens, she and Tony were left hanging upside down with their backs bent over for more than an hour when their roller coaster became stuck. The cleaning work and the Busch Gardens incident combined to cause several of Gina's discs to herniate. In 1993 one of those herniated discs ruptured and impacted her spine, causing it to collapse. She underwent five complicated back surgeries at the University of Pennsylvania Medical Center and was bedridden for a year. Eventually, after a year of surgery and bed rest and supported by her devoted husband and two children, Gina was able to walk again with the help of a cane. But her back was never the same, and she was unable to work again. For Tony, his mother's resilience was inspirational.

Like the famous Levittown on Long Island, New York, Fairless Hills in Pennsylvania was a product of the enormous American suburban housing boom of the early 1950s. As historian James T. Patterson has suggested, suburbs such as Levittown and Fairless Hills were made

possible by the enormous increase in car ownership.[1] Fueled by generous mortgages from the Federal Housing Authority and the Veterans
Administration and an almost insatiable demand for housing from
young World War II veterans and their large families, these industrial-
scale suburbs grew exponentially. In Bucks County, Pennsylvania,
renowned developer William Levitt even built a second Levittown
not far from Fairless Hills where Tony grew up.

Fairless Hills was a smaller Levittown, a typical example of the
housing of the period: prefabricated mass-produced homes built on
small lots and sturdy, good-quality buildings with modern home appli-
ances. What was unusual about Fairless Hills was that it began as a company town. Built in 1951 and financed by U.S. Steel, the development
was named in honor of Benjamin Fairless, president of the steel giant.
Fairless Hills was primarily intended to house U.S. Steel workers at its
Fairless works steel plant, although the development was also open to
nonemployees.

Opened in 1951, U.S. Steel's massive open-hearth Fairless works
was the biggest and most efficient steel mill of its kind in the United
States. With nine smokestacks stretching like enormous metal fingers
into the Pennsylvania sky, Fairless personified the strength of the post-
war U.S. industrial economy. And for thirty years, the steel works gave
the people of Fairless a good slice of the American dream, enabling
them to live comfortably and put their children on the escalator of
upward mobility. By 1973, Fairless employed ten thousand people. But
as Tony was growing up, a whole way of life ended in Fairless and in
many other American steel towns. In 1991 when Tony was eight, the
giant Fairless steel mill shut down. This was a familiar story in America's steel towns brought about by blinkered, complacent management;
antiquated technology; and lower-cost Japanese competition.

All around the Fuscellaros, homes had to be sold to pay back taxes,
and some marriages ended. But the Fuscellaros were devout Catholics
whose strong religious faith helped sustain them through this eco-
nomic and social collapse. As Anthony recalled, "We chose not to leave;
we chose to hold on to each other." Gina added, "It was a place to call

home. Everyone was well rooted in the town, in each other, and in their faith. Faith really tied us together." All the Christian churches pulled together, ensuring that every family had adequate food. The state and local governments also worked hard together to ameliorate the worst of the crisis. Above all, there was an extraordinary community effort. Neighbors who still had jobs created jobs—mowing lawns, washing clothes, whatever it took—for neighbors who had lost theirs. Tony's father worked for the State of New Jersey and, unlike most of his neighbors, still had a job, so the Fuscellaros were in the vanguard of the community's response. Gina and Anthony's selfless leadership gave Tony a master class about the importance of serving others in a time of need.

Gina was a devoted mother who not only ran the household but also set a high example of what a selfless, moral person looks and acts like. Anthony took on additional jobs to support his family and pay for their education in Catholic schools. Throughout Tony's childhood and adolescence, Anthony was a model of resilience, determination, steadfastness, and selfless devotion to his family and community. He always told his son "never give in, never give up."

Anthony and Gina's influence can be seen in Tony's intense religiosity and unusually strong moral character. The Fuscellaro family was active in the St. Frances Cabrini Church and engaged daily with the school, the church, or its parishioners. Anthony and Gina set a great example for Tony and Alicia not only as devoted parents but also as devout Christians and responsible, moral, people driven to volunteer to support their church in a wide variety of roles. Tony followed their example, volunteering as a server and a youth tutor.

At home, Gina and Anthony set Tony a clear moral code based on a few basic principles: obey God and his commandments; always do the harder right; never lie, steal, or cheat; never ignore someone who needs help; and never lose an opportunity to serve others. Gina recalled that Tony's "father showed him how to be a loving member of society." A good example of this code in action occurred when Tony was nine and brought home a church-sponsored sacrifice project. The question was what he would be willing to give up for others. He offered a favorite toy soldier. His father said that was fine, but why not give up something he loved even more: the gift of sight and the joy of reading. Tony agreed and put eyeshades over his eyes for a week in the belief that somewhere through surgery God would restore the sight of someone who had lost it.

Gina and Anthony always believed that the greatest gift you can give your child is a first-class education. To pay for it, Tony's parents worked incredibly hard. Growing up, it was not uncommon for Tony to see his parents work multiple jobs to pay the tuition at St. Frances Cabrini Grade School and later the elite Holy Ghost Preparatory School. Many years later, Tony summarized their influence: "I was always impressed and inspired and had to make sure that I carried my own weight."

Both schools left an imprint on Tony by reinforcing the sound values his parents had taught him. His grade school was one of those founded by Frances Cabrini, the first American to be canonized as a saint by the Catholic Church. She was a devout force of nature: a fiercely determined Italian American nun with unshakable faith in God who founded sixty-seven schools, hospitals, and orphanages throughout the United States in the late nineteenth and early twentieth centuries. The schools' value-centered education emphasized faith as well as selfless service to others. For Tony, this approach to life was powerfully reinforced by the Holy Ghost Prep, founded in 1897 by members of the Spiritans, the congregation of the Holy Ghost. Tracing their roots back to France in 1803, the Spiritans have historically accepted difficult missions. In its impressive gray granite buildings, the elite prep school emphasized faith, prayer, and good works. This was all good preparation for West Point.

When Tony was a boy, his intellectual gifts were clear. He earned excellent grades at St. Frances Cabrini Grade School and later at Holy Ghost Prep, one of the leading private high schools in Pennsylvania. There he maintained a B+ average in a demanding college preparatory curriculum and earned the admiration and respect of his teachers, who praised his disciplined systematic approach to learning. They respected his desire to learn, improve, and grow. As his drama teacher wrote, Tony was the student "who was always prepared, who had fully absorbed the readings, thought about his character and turned the thinking into believable acting."[2] The teacher added that he had come to respect Tony's "keen perceptions into character motivation and subject" as well as his "ability to translate analysis into performance." In debate, Tony had to learn how to develop clear, concise arguments for or against a resolution and present them persuasively in front of judges. As a high school sophomore and junior, Tony was a state finalist ranking in the top six students in Pennsylvania and a national semifinalist. Observing Tony's performance as part of Holy Ghost Prep's nationally ranked debate team, his literature teacher, Jeffrey Danilak, described what he saw as Tony's "uncanny ability to analyze and present the main points of his argument" that "left the other team speechless."[3]

If there was one extracurricular activity that played a central role in Tony's development as a leader it was his captaincy of Holy Ghost Prep's storied Forensics Society. This society focused on preparing students for competition at the regional, state, and national levels in speech, debate, and drama. Its goal was to give students a platform to find their voices, hone their delivery, and learn to advocate for a better society. Under the dedicated, dynamic leadership of Tony Figliola, Holy Ghost Prep's Forensic Society had won numerous championships at the state and national levels. The Forensic Society gave Tony the opportunity to develop as a confident public speaker, a creative and critical thinker, a writer, a skilled debater, and a leader. His principal focus was as a dual performer. He and a partner prepared a ten-minute scene from a play and then performed it in front of an audience. To do this, Tony had to read the whole play and select parts of it so that combined together, they could accurately represent the essence of the play as a whole. As a junior, he played a Jewish attorney representing a Neo-Nazi skinhead accused of murder in the play *Cherry Docs*, an intense and brutally frank look at hatred and what it takes to eradicate it through mutual education. As a senior in another play, Tony played an older brother struggling to cope with the emotional challenges of looking after his deaf younger sibling. Both plays embodied the Forensic Society's main goal of using serious drama to make people think about bettering society and required an exceptional level of empathy, insight, and understanding from a teenager.

Tony was a born leader. As a junior, he cut an impressive figure: impeccably dressed, hair perfectly combed, articulate, self-assured but never arrogant. He was also someone who commanded respect not only because of his commanding presence but also because of his unquestionable integrity. Not surprisingly, as a junior and senior he was elected captain of the Forensics Society, where he led by example. As one of his teachers wrote of him, "When kids see him working, they work. When they see him focusing, they focus. When they notice him carrying himself with confidence and acting as a mature leader, they decide it is silly to act silly. When other officers see him do lots of organizational work, they feel guilty not having done their lot, and wind up contributing even more."[4]

And if all this was not enough, Tony was elected his class representative in Holy Ghost Prep's student government, played varsity baseball, and served as moderator of the school's community service program for students in ninth through twelfth grades.

If Tony was an inspirational leader among his peers in the Forensic Society, he was also a caring, compassionate young man who devoted an unusual amount of time to selfless service to the hungry, the homeless, and the physically disabled. Beginning as a high school freshman, Tony volunteered four hours a week as a tutor at St. Francis Cabrini Grade School and, according to his teacher Margaret Flynn, "helped make this program a success."[5] Not content with this, Tony served meals twice a week at the St. Francis Inn, a soup kitchen for the poor, and served as a volunteer soccer coach for the Special Olympics. Here he showed great patience and caring. As if this were not enough, in his junior and senior year Tony served as community service outreach coordinator for Holy Ghost Prep and its branch of the National Honor Society. And there was plenty to do. His school had mandatory community service hours for each class year, and the National Honor Society also had a large community service component. Tony led the program and helped connect Holy Ghost Prep students with community service projects in the local area.

What stands out in Tony's West Point application is his deeply felt humanitarianism: his concern for the well-being of people, especially his fellow students. In answer to a question about the social issue in his community that most concerned him, Tony wrote about gun violence in schools. He applauded increased security measures but criticized their ineffectiveness. As he saw it, the real problem was that the school authorities could not and the students would not identify potential shooters.

Imbued by his parents with a deep sense of duty and commitment to serve others, Tony grew up with a deep respect for public servants, civilian and military. He was especially drawn to the sense of higher service provided by the military. As he recalled many years later, "I loved the idea of defending my country and defending those who could not defend themselves." And he was intrigued by the challenge of getting accepted and succeeding at West Point.

In earning a place at West Point and building a distinguished military career, Tony was plowing a lonely furrow. To begin with, unlike most West Point cadets, there was no long tradition of military service in the Fuscellaro family, and there were no distinguished USMA graduates. However, both of Tony's paternal grandfathers served in the Army in Korea. As they saw it, they had fulfilled their service obligation to the nation, no more and no less. They never became advocates for military service. Only Gina's oldest brother, Joe Deering, became a professional soldier. Throughout the 1980s and 1990s Tony's uncle Joe had served in the Army, spending a significant amount of time at Fort Bragg, North Carolina, home of the 82nd Airborne Division. He loved the Army, and his enthusiasm was infectious. Uncle Joe gave Tony photographs of parachute jumps. One Christmas, he gave Tony the Iron Mike, the traditional Fort Bragg paratrooper statue that Tony kept in his room until he went to West Point. Because Tony was seriously interested in serving his country and because of his obvious intelligence and leadership potential, Uncle Joe steered him toward West Point.

But Tony's guidance counselor at Holy Ghost Prep was not supportive. The school was known for its ability to produce outstanding doctors and lawyers, not military officers. When Tony applied to West Point in 2000, the last Holy Ghost Prep student who had applied to the USMA was back in 1982, and although he had got ten in, he had failed to complete the plebe year. That student's experience had convinced Tony's guidance counselor that if he could not succeed at West Point, then no one else from Holy Ghost Prep could either. The guidance counselor even told Tony that applying to West Point was a waste of time and effort and that he would never complete the four years there. The service academies were too difficult to get into and too stressful to succeed in. Instead, he and other guidance counselors usually steered students toward the Ivy League, where the majority of Holy Ghost Prep students earned places and succeeded academically.

Even Tony's own parents had concerns about his application to West Point. His parents knew he was smart enough but wanted to be sure that this was something Tony really wanted. His father had two primary concerns. The first was that he knew that being an Army officer was a tough profession that demanded an unusual level of mental and physical toughness. The second concern was that the U.S. Army was downsizing after the end of the Cold War and might not offer Tony a long-term career. Anthony Fuscellaro was not being negative: he simply wanted his son to understand the harsh realities of military life and to be absolutely sure this was what he really wanted to do with his life. Once his parents were satisfied that going to West Point was something Tony really wanted, they embraced his desire. His mother acted as his guidance counselor, steering him through the long West Point application process and ensuring that he never missed a deadline.

During the application process, Tony visited West Point once and stayed overnight with a plebe. Tony also applied to West Point's summer academic workshop but was not accepted. That disappointment merely intensified his determination to succeed. In December 2000, he received early-action acceptance for West Point. He then competed for and secured a formal nomination from Congressman Jim Greenwood, who represented Pennsylvania's Eighth Congressional District.

But proving a point to his parents and his school counselors was not the only reason Tony applied to West Point. Like the other elite souls, moral reasons also played an important role in motivating Tony toward a career in military service. He had attended St. Frances Cabrini Grade School in Fairless Hill, where he and his family were heavily involved in church activity. Like Ross Pixler, Tony was also an altar server and was active in the youth ministry. To distinguish himself from other West Point applicants, he emphasized his dedication to serving others in his application to the USMA: "I feel that my experience in many service activities is a major characteristic of my life. I participate in things from soup kitchens to grade school tutoring to the Special Olympics. My attendance at a Catholic Grade School and then a Private High School have instilled a characteristic of service in me, along with a firm belief in God. Most men my age do not have the relationship with God that I do, and that along with my dedication to community distinguishes me as an individual."[6]

Finally, Tony was "always a big team-building type of guy," as he put it, and he thought that this was more present in the Army than in the Navy and the Air Force. Besides, he loved the idea of the airborne ranger or infantry officer in the field leading a team. Boats and ships never excited him.

The irony is that the extended Fuscellaro family, which had no real tradition of military service and had never planned to establish one, has now became a military family. Today, Tony and four of his first cousins are on active service across all branches of the U.S. armed services.

3

ROSS PIXLER

Ross Pixler grew up to be a striking figure: five feet, eleven inches, trim with short, light-brown hair, crystal blue eyes, and the luminous smile of a young Robert Redford. Ross was a son of the West: born in Montrose, Colorado, he was raised in Phoenix, Arizona. He was a born soldier and, as a little boy, watched war movies and read books about military history. In high school, Ross was the battalion commander of the district Junior Reserve Officers' Training Corps (JROTC). Later he wrote that "clergy members . . . are always talking about a 'calling.' I felt I had a calling too . . . but it was to a different service. It seemed like everything in my life was leading me towards the Army. I knew very little about West Point, but I felt compelled to go there."[1]

Ross Pixler was born in August 1982 in Montrose, Colorado, the youngest of three children. His family moved to Phoenix, Arizona, when his father was appointed assistant U.S. attorney there in 1989. Reid Pixler, Ross's father, a respected lawyer, was the kind of exemplary patriotic public servant America's Founders hoped for: a man of principle, exceptional courage, unquestionable integrity, and devotion to public service. In that selfless spirit, Reid even volunteered for the U.S. State Department's Rule of Law Program in Iraq, where he worked continuously for twenty-one months helping the Iraqi government build a new criminal justice system.

Ross's mother, Larissa, was equally selfless, devoted to her family and public service. Born in Stamford, Connecticut, Larissa was a natural linguist with a flair for Spanish, which began with her childhood (and lifelong) friendship with the youngest daughter of Mexican business associates of her family. Larissa studied Spanish in high school and at Southern Connecticut State College. As a junior, she enrolled in Adams State College, Colorado, so she could access its exchange program with the University of the Americas in Puebla, Mexico. Later, after they married, Larissa moved with Reid to Phoenix. In July 1989, federal law enforcement urgently needed Spanish wiretap and document translators. Larissa embraced this challenge and soon became one of the most sought-after translators in federal and state law enforcement in Arizona.

As assistant U.S. attorney, Reid Pixler earned an enviable reputation as a courageous, innovative prosecutor who tracked down and confiscated the financial assets of the Colombian and Mexican drug cartels. In one case, for example, Reid invoked the U.S.-UK Mutual Legal Assistance Treaty, successfully grounding a fleet of drug-carrying aircraft by seizing the unearned insurance premiums from Lloyds of London. Inevitably, his success against the drug barons came at a price. A detective on his team received a credible death threat, and Reid himself lived under real and continual nervous strain, worried that the drug cartels might target him, Larissa, or their three children. This concern was heightened because Larissa, as a Spanish-language wiretap translator for law enforcement agencies, could regularly hear drug traffickers say how much they hated Reid. It was a chilling experience for Larissa, but neither she nor Reid ever wavered in their deep commitment to upholding the rule of law and stopping the drug traffickers.

At home, Reid and Larissa had to have life-and-death discussions with Ross, his older brother, Ryan, and older sister, Kelley. These included warning their children never to fraternize with anyone of any age they didn't know lest they be targeted by assassins from the Colombian or Mexican drug cartels. Add in their parental admonition that anything less than strict compliance with the rule of law was unacceptable, and you have a recipe for an unusual childhood. But it was an inspiring one: two warm, loving parents who dedicated their lives to protecting other families from the inflow of dangerous drugs smuggled in and sold by some of the most evil criminals imaginable. It was a noble and important cause greater than themselves.

Devoted to their children, Reid and Larissa gave them every possible opportunity to grow and succeed. Ross later wrote of them that "they live most of their lives in committed servitude to make the lives of their children better. There is no one else in the world I could admire more for their constant dedication to putting others first."[2]

In the 1990s Phoenix was booming, the fastest-growing metropolitan region in the United States. Ross and his family lived in Ahwatukee, a large suburban area that forms the southeastern boundary of greater Phoenix. Nestled in rugged, rolling foothills, Ahwatukee's subdivisions were considered some of the most desirable neighborhoods in greater Phoenix: sturdy, well-built, attractive homes as well as good schools and safe streets. Ahwatukee was generally quiet and safe, but in 1994 Ross and his family were shocked when a boy Ross walked to school with murdered another seventh grader in that boy's home after school.

Crime struck even closer to home when Ross was attacked by knife-wielding thugs twice in his boyhood. On both occasions, Ross responded with courage and appropriate force. He was not a boy you threatened. The first incident occurred in a Barnes & Noble bookstore when a young man demanded he leave the store. Ross refused, fearing for the safety of his girlfriend. He protected her by standing between her and the knife-carrying assailant. His older brother approached the young thug from behind and subdued him. The second incident—a case of road rage—occurred when Ross was sixteen. Here is how he recalled it:

> I was on my way home from working at Senator John McCain's office and five teenagers jumped out of a jeep in front of me and came racing back to the driver side door of my Pathfinder. I kicked the door open so hard at the right time, that I laid the first guy

out on the ground, put a staple gun to his forehead and scared his friends so bad that they immediately turned around and ran away. They all thought my staple gun was a real gun because it was chrome and flashy. I had been using it to put up yard signs for the campaign. It was all I had to defend myself, but it worked.

At first, Ross and his family lived in the Lakewood subdivision in a two-story, wood frame corner house with stucco siding. In 1995, they moved two miles away to a large four-bedroom home of similar design in the Mountain Park Ranch area. Both homes had swimming pools in small back gardens and were close to schools.

As a little boy, Ross was an irrepressible, fearless, "adventure seeker." On one occasion, his older brother, Ryan made a rope escape ladder, tied it to their bunk bed, and dropped it out of their bedroom window. Ryan asked Ross, the youngest and lightest, to try it out. He did, but he fell off the ladder to the garden about twelve feet below. Undaunted, Ross climbed back up the rope ladder into the bedroom. On another occasion, Reid took Ross skiing. Reid had taught him to ski to a beginner's level, but these trails were confusing, and Reid made a wrong turn. He and Ross accidentally turned onto the slope intended for advanced experienced skiers. Without missing a beat, Ross skied down the steep slope. When Reid followed him, Ross was waiting at the bottom. With a cheeky grin on his face, he said, "C'mon, Dad. Hurry up!"

Ross's growth as a selfless moral leader was reinforced by his and his family's active participation in the Corpus Christi Catholic Church, the Boy Scouts, karate, and JROTC. Originally founded in 1988 in a temporary church, the Corpus Christi Catholic Church grew so rapidly that the Catholic archdiocese had to build a new permanent church, which was completed in December 1996. Built of brown cast concrete, with a life-size sculpture of Jesus giving communion outside, and a two-story tower in the old Spanish missionary style, the new church seated seven hundred congregants. There, Ross was a volunteer, an usher, and an altar server under his spiritual mentor Father Lewis Sigman, the church's founding pastor. Father Sigman was a committed charismatic clergyman of great integrity and moral force. Guided by Father Sigman, Ross not only attended church classes and was active in the youth group but also pursued religious awards that he could receive through Boy Scouts. Not surprisingly, Ross won every award available. Father Sigman and other dedicated clergy were important in the formation of Ross's religious faith.

Ross was a fiercely stubborn young adventurer who loved the outdoors. He thrived in Boy Scouts, which not only gave him outdoor outlets for his restless energy but also cemented the sound values and principles Reid and Larissa had taught him: duty, honor, selflessness, and respect for the rule of law. By chance, most of the Scout troop's adult leaders were engineers who gave the boys great exposure to the basics of engineering and construction. Reid, who was also a volunteer Scout leader, complemented the troop's engineering education with visits to federal and state law enforcement agencies. On one occasion, Reid arranged for the boys to observe the federal trial of a Mexican drug trafficker. Reid's influence and that of his older son, Ryan, influenced Ross's choice of Eagle Scout project: mobile targets at the greater Phoenix federal law enforcement firing range. For his Eagle Scout project, Ryan had built mobile targets on sleds. Three years later, Ross restored them and put them on wheels, earning his Eagle Scout rank at age sixteen.

Karate also proved valuable. For ten years, Ross studied the art and science of martial arts with the best: sensei Rick Savagian, a student of Shihan Toshio Osaka, an eighth-degree black belt, founder of the International Karate-Do Center, and chairman of the USA Wado Ryu. Savagian trained Ross in the core tenets of Japanese karate "to value and to understand the three standards of honor: obligation, justice and courage." Karate helped Ross develop remarkable physical and mental resilience. Ross also saved a woman who was being attacked just outside the karate school. Coming out from karate class with his older brother, Ryan, they saw a man beating a woman in the parking lot. They immediately ran to her rescue. Ryan threatened her assailant, saying, "If you don't stop hurting this woman, I will hurt you, but [gesturing to Ross] he will kill you." The man ran off. Ross would never have killed anyone, but the intensity of his moral outrage was palpable and intimidating. At age seventeen, Ross earned a black belt in karate, its highest rank, and regularly stood up for high school students who were smaller or getting bullied.

The JROTC was the fourth and final influence on the development of Ross's remarkable moral character. At age fourteen, he discovered the JROTC thanks to his family's neighbor, Chris Narmi, who gave him material about the program and encouraged him to join. Chris's uncle, Rear Adm. Ronald Narmi USN (Ret.), a regular visitor to Phoenix, was a major influence on Ross joining the JROTC and pursuing a military career. So, when his parents took him to Mountain Pointe High School to help him identify which student clubs might interest him, he was already thinking of the JROTC. At the school event, Ross met two JROTC cadets in uniform, and as he later told me, their "exceptionalism" appealed to him. As a boy, Ross had always dreamed of being a soldier. Here was an opportunity to start. One of the two JROTC cadets was Jared Sibbitt, who would become a lifelong friend. Ross served under him and then succeeded Jared as JROTC battalion commander.

Jared became not only Ross's best friend but also his role model. Jared was a charismatic and highly articulate young man, compassionate and caring. Like Ross, Jared was a fierce adventure seeker. Despite suffering from diabetes, Jared was also a fearless football player, athlete, and mountain climber. As Junior ROTC battalion commander for a district unit of about two hundred high school students, Jared proved himself to be an excellent servant-leader, someone who listened and gave good advice, spurring other students on to greater achievement. If Jared had a weakness, it was that like Ross he could be overconfident in his ability to tackle any problem.

For an intellectually gifted, energetic adventure seeker, the JROTC was a perfect fit. Founded in 1916 as an Army program, the Lyndon Johnson administration and Congress redesigned the JROTC in 1964 to embrace all branches of the U.S. military. It was and is a character-building and citizenship education program taught by retired officers from all of the armed services. Some were and are former active duty officers, former reservists, or national guardsmen. Ross's JROTC program was based at nearby Marcos DeNiza High School. His instructor was Lt. Col. Peter Stolze (Ret.), a 1971 ROTC graduate of Arizona State University known to his friends as "Pat." Lieutenant Colonel Stolze had served as a cavalry officer in the days before aviation became a separate Army branch. As a result, he served both as an armor officer and a Huey pilot. His Army career took him to two of the Cold War's frontiers: West Germany and South Korea. For Ross, Stolze was a demanding but always caring and supportive JROTC instructor and soon became a close friend of Ross's family.

The JROTC was a demanding program, with classes on leadership, citizenship and government, military history, physical training, and outdoor activities. For Ross this meant classes starting at 7:10 a.m. on Mondays, Tuesdays, Thursdays, and Fridays with uniform and kit inspection on Wednesdays. And when he became the battalion commander, it meant returning to the JROTC office at Marcos DeNiza High School in the late afternoon to take care of his unit's paperwork.

Very quickly, Lieutenant Colonel Stolze spotted Ross's excellent leadership qualities. As he later told me, "Ross led from the front. He was always so positive, always so engaged, a great speaker who always turned out a quality product."

Lieutenant Colonel Stolze added that Ross always showed him "great respect" and was a "meticulous planner" who implemented those plans with energy and effectiveness. Under Ross's leadership, the JROTC unit based at Marcos DeNiza High School won every trophy they competed for, a record never equaled since. Ross himself was named the top JROTC candidate in Arizona.

As Ross wrote in his application to West Point in December 2000, becoming the Junior ROTC commander was not only his greatest achievement but also a position in which he learned so much. "I have worked towards this goal for four years, and I have contributed thousands of hours in training to become a better citizen and cadet. Every day since I assumed my duties, I have learned so much about leadership, discipline, friendships and understanding. There are many sacrifices that come with a leadership position. Being responsible for 168 cadets from 6 different schools has caused me many sleepless nights, and prepared me for life's many trials and adversities."[3]

As a child and as a teenager, Ross was surrounded by selflessness and excellent role models. Academically, however, he was at first a mediocre student in the small private Catholic school he attended until seventh grade. When he transferred to Mountain Pointe High School he began to excel, earning straight A's. Ross had decided to "get real," to work hard academically and succeed.

Unlike three of the other elite souls, Ross had a significant family military tradition. His great-grandfather served in the U.S. Navy in World War I. Quartermaster on a transport ship on the transatlantic convoy run, he had a lucky escape from early death. Enjoying the English countryside on shore leave, he was late back to his ship, which mysteriously disappeared in the Atlantic and was never heard from again. Both of Ross's grandfathers served in World War II. Larissa's father had served in the Army under Gen. George Patton, where he was put in charge of food distribution at the Dachau concentration camp after his unit liberated it. He suffered PTSD for decades afterward because of the horrors he witnessed at Dachau. Reid's father, a graduate of the Colorado School of Mines, was a highly decorated transport pilot for the U.S. Army Air Corps in the Pacific theater. Larissa's older brother served in the Army in Vietnam, her brother-in-law served in the Air Force, and her cousin served in the Army, retiring as a lieutenant colonel. Ross's uncle served in the Air Force and also retired as a lieutenant colonel from the U.S. Air Force Academy faculty. Within Ross's family, there was thus a push toward the Air Force Academy. Ross resisted fiercely because the Army appealed more to him. He did not want to be a pilot because he felt it was disconnected from leading soldiers. He wanted to lead people, not equipment.

As with the other elite souls, moral reasons were important in driving Ross to West Point and to serve as a military leader. For him, it was his way to deal with what he saw as the crisis of morality in America today. In his application to West Point, he wrote that "there are too many people, old and young, who either do not know the difference between right and wrong or do not care. There are not enough people who are willing to step forward and take responsibility. Decision-making is a process that no longer emphasizes values and morals. Money and self-gratification have taken the place of ethics and morality."[4]

But there was a major obstacle on the road to West Point. Throughout his childhood, Ross had suffered repeated asthma-like attacks that sent him to the hospital emergency room. In fact, Ross never had asthma. His wheezing was caused by his lungs' reaction to environmental allergens as well as the smoke from wood-burning fires in the area trapped by temperature inversion. When he was eighteen months old and in the emergency room's plastic oxygen tent, he climbed out of the tent and sat on it while waiting for a nurse to come. Remarkably, Ross learned to overcome these attacks by continuing to run through them. He likened it to running while wearing a gas mask or breathing through a straw: it was difficult but not impossible. Remarkably, despite this impaired breathing, Ross could run a mile in under six minutes and thirty seconds. When Ross was sixteen his mother suggested that he go to a pulmonologist, whose tests confirmed that he did not suffer from asthma. But for the Army medical board this was not enough, and Ross's initial application to West Point was denied. It took an act of exceptional selfless courage in a snowstorm on Mount Orizaba in Mexico to change their minds.

In late December 2000, Ross and Jared set out to climb "the unforgiving mountain," Mount Pico de Orizaba, the tallest mountain in Mexico.[5] They were experienced mountaineers, but this expedition went badly wrong. Battered by severe winter storms with driving winds, low visibility, hail, and heavy snow, the close friends became lost high up Orizaba. For two nights amid the snow-packed peak, Ross and Jared endured temperatures up to 60 degrees below zero. They took turns staying awake so they would not freeze to death. But early on New Year's morning 2001, Jared ran out of insulin. He told Ross that if he could eat a lot of snow, he could survive for ten hours without it. It was up to Ross to save his friend's life. Every minute counted; the clock was ticking.

Ross ran ten miles through mountainous terrain to find help. Sleep-deprived, he had not eaten for two days and nights. Ross was also suffering from severe dehydration, frostbite, hypothermia, exhaustion, muscle cramps in his legs, and a sprained ankle. At one point his vision was blurred and he began to hallucinate, but he kept going and, with the help of a family of mountain villagers, found medical help and a rescue team for Jared. It was an act of extraordinary selfless courage. Once Lt. Col. Peter Stolze, Ross's JROTC instructor, learned that Ross had been denied a place at West Point, he immediately wrote a letter to the USMA's Admissions Committee requesting a reversal of its decision. In his letter, Lieutenant Colonel Stolze told the story of the dramatic rescue that Ross had performed on Mount Orizaba, adding that anyone who was even a mild asthmatic could not possibly have done what he did. Reid too was active. He spoke to the USMA's military liaison officer in Phoenix, Lt. Col. Jack Ruffing (Ret.), explaining the full story of how Ross had saved Jared Sibbitt's life atop Mount Orizaba. Once in possession of these new facts, Lieutenant Colonel Ruffing telephoned his old West Point roommate, Brig. Gen. Eric Olson, then commandant of cadets. Brigadier General Olson immediately intervened on Ross's behalf, and the Admissions Committee reversed its earlier decision and gave Ross a place in the new class entering the USMA in June 2001. Ironically, at West Point Ross earned the Best Physical Score award in his year group.

4

BOBBY SICKLER

Bobby Sickler grew up to be an impressive figure: five feet, eleven inches, trim with a high, prominent forehead, striking blue eyes, and a wide, welcoming smile. He was the son of a distinguished U.S. Marine colonel and spent his childhood on Marine bases in the United States and Japan until his father retired and took over a farm in West Virginia that had been in his family for three generations. There in Pendleton County in the Appalachian Mountains, "in the middle of nowhere" as Bobby put it, he spent his teenage years. He was the heir to a distinguished family heritage of military service as citizen-soldiers dating back to the American Civil War. From the time he was twelve years old, Bobby had wanted to be a career military officer like his father and worked single-mindedly toward earning a place in one of America's service academies. As a boy and as a teenager, Bobby was kind, unusually calm, humble, exceptionally intelligent, highly motivated, and self-disciplined. In the words of his West Point interviewer, he was "a Find."[1]

Bobby was born on October 24, 1983, in the U.S. Army hospital in Fort Belvoir, Virginia. His father was serving on the Naval Air Staff in the Pentagon. As a young boy, Bobby grew up in post housing on a succession of Marine Corps bases. In 1995, however, after a twenty-six-year career in the Marine Corps, Bob Sickler retired to devote more of his time to his family and to rebuild the farm that had been in his family for more than a century but had been uninhabited for twenty years. To that end, Bob and his wife, Patty, moved their family to rural Pendleton County, West Virginia, just off U.S. Route 33.

Bobby's parents were model public servants: a Marine Corps officer and a federal investigator turned public school teacher, respectively. They were the kind the Founders hoped for: two people of high moral principle, unimpeachable integrity, and dedication to the well-being of the nation and their family.

Bob was born in Harrisonburg, Virginia, and grew up in Tunkhannock, a small town in northeastern Pennsylvania just over thirty miles from Wilkes-Barr. He was educated in Missouri at Saint Louis University's Parks College of Aeronautical Technology, graduating in 1969 with a BSc. Later during his military career, Bob earned a master's degree in international relations from the U.S. Naval War College, Newport, Rhode Island, and an MBA from Salve Regina University in Newport. When he was drafted in 1969 he volunteered for the U.S. Marine Corps, where he made a twenty-six–year career.

Bobby's mother, Patty, was born in Saint Louis, Missouri. Her father, a World War II veteran, was a Marine master sergeant on recruiting duty when she was born. Patty grew up in Cincinnati, Missouri. She earned a BS in education from Tennessee Tech University and a master's degree as reading specialist from Marshall University. She met Bob in 1979 in Tennessee when she was a federal investigator with the U.S. Postal Service and Bob was attending a military school in Millington, Tennessee. Patty was a homemaker until Bobby and his younger sister went to school. Thereafter, Patty taught in Yuma, Arizona, and in Department of Defense schools in Okinawa, Japan, and Quantico, Virginia. When Bob retired to West Virginia from the U.S. Marine Corps, Patty taught in the Pendleton County schools until 2017.

Pendleton is West Virginia's biggest county but has the fifth-lowest population in the state. It is a tale of two counties. The first is strikingly beautiful: rugged, densely forested mountains and lush green valleys sitting at the head of the South Branch of the Potomac. The second Pendleton is distressed and so typical of modern rural America: declining population and a struggling economy. Within Pendleton County Bobby spent his teenage years in Brandywine, an even older and poorer community where median family income was far below the national average and where the population had been falling at an even faster rate than the rest of the county.

Here in this beautiful but distressed rural area, Bobby spent his teenage years. He grew to manhood in a two-story, three-bedroom wood frame house built in 1902. The farm had been owned by the Sickler family for generations. Bobby's great-grandfather had bought the land in 1896, but by the time he and his family moved there in 1995, the farm had lain dormant, uninhabited for twenty years. Reclaiming the farm from nature took five years. Then and now small farming was not very profitable, but that was fine with the Sicklers. Bob had his Marine Corps pension, and Patty had her teacher's salary. The modest profits from the farm helped supplement their income.

As a boy growing up, Bobby had a very close relationship with his father, who was an especially important influence. Every evening after school and every weekend, father and son worked together on the farm. Later, Bobby described it to me as "one-on-one tutoring in values and service." Bob and Patty demanded that Bobby and his sister live by a strict moral code. They taught them moral character through focused discussions about each person's moral obligations in life, using examples of good and bad behavior. They reinforced this by regularly attending church services at nearby Calvary Lutheran, a beautiful Victorian-style redbrick building with a two-story bell and a clock tower entryway.

As a boy, it was clear that Bobby was a gifted, intellectually ambitious student. He was always observant, engaged, and curious. Like Stephen Tangen, Bobby was the kind of "questioning student" educators love: the boy who always wanted to learn more than was offered in the course or required for the test. According to his teachers, Bobby's great strengths were in math and science. Precalculus was the course that interested him most. It was not offered at Pendleton High School, so he convinced the high school math teacher to allow him to take precalculus as an independent study. Bobby was also well rounded. Although math and science were his greatest strengths, he worked hard to ensure that his grades in English were equally good. He also excelled in public speaking and won several awards for oratory.

In her letter of recommendation in support of Bobby's application to West Point, Marsha Keller, his high school guidance counselor, stated that "Bobby is an outstanding academic student who is highly motivated, well organized and extremely responsible.... He is a self-starter and is self-disciplined."[2]

Unlike Nick Eslinger, Tony Fuscellano, and Stephen Tangen, Bobby was heir to a long and distinguished family heritage of military service. Generations of Sicklers had served as citizen-soldiers dating back to Bobby's great-great grandfather, a wounded veteran of the American Civil War shot on the eve of the of the Battle of New Market (May 15, 1864) in the Shenandoah Valley. In World War I, Bobby's great-great-uncle Earl served with the U.S. Expeditionary Force in Europe, where he was gassed by the Germans. Patty's father had joined the U.S. Marine Corps in the mid-1930s and served throughout World War II in the Pacific theater, mustering out as a master sergeant. He had fought in the Battle of Iwo Jima and witnessed his fellow Marines raising the American flag over the Japanese island in that now iconic moment. Bobby's paternal grandfather had served in the U.S. Army during World War II and was badly wounded in the Battle of the Bulge (December 1944–January 1945). He spent the rest of the war in Britain recovering from his wounds.

But Bobby's greatest influence was his father, the family's one and only professional soldier. In 1969 before he was drafted and joined the U.S. Marines, Bob had graduated with a pilot's license from the Parks College of Aviation and Engineering at Saint Louis University. During the Vietnam War most Marine Corps draftees chose to leave after their war service was over, but not Bob. He chose to stay for twenty-six years. Unusually, for a young Marine officer in the late 1960s, Bob did not serve in Vietnam. At Quantico, his was the first Officer Candidate

School class where only half were deployed to Vietnam. Bob was disappointed. He was well trained and well prepared, but his orders changed close to deployment.

At the beginning of his Marine Corps career, Bob was an armor officer but, having studied aeronautical engineering, became an aviation maintenance officer. Later he took up civilian flying as a hobby and became a flight instructor. He would imbue his son with his love of flying and the U.S. military. Bob remembers giving the controls of a Cessna 110 to a four-year-old Bobby, his first flight. Like his father, Bobby fell in love with flying and really wanted to be a Marine Corps or Navy pilot but had to abandon his goal because of poor eyesight. The laser eye surgery that would have corrected his vision and enabled him to fly was not yet available.

From a young age, Bobby never wanted to be anything else but a military officer. But as a child he lacked self-confidence, something his father identified and was determined to remedy. When Bob was deployed in Okinawa, he signed nine-year-old Bobby up for karate lessons. His teacher, Katherine Lokadopulous, an expatriate American, took a personal interest in Bobby and launched him on the path to earning a black belt at age seventeen. She did this in two ways. The first was by educating Bobby in the core tenets of Japanese karate "to value and to understand the three standards of honor: obligation, justice and courage." The second way Lokadopulous did this was by helping Bobby develop remarkable physical and mental resilience. As he later wrote in his West Point application, "She made us work really hard and I liked that she had us doing knuckle pushups on the concrete. I got to where I could do about a hundred pretty fast."[3]

Another step Bobby's father took when he retired from the U.S. Marine Corps and settled in West Virginia was to give Bobby a target pistol and teach him to shoot. Bobby's marksmanship continued to develop, and he learned to use a rifle. On one occasion on the farm, he shot a moving deer at 225 yards. This high level of accuracy would later save American lives in Iraq. Finally, to continue building Bobby's self-confidence, Bob encouraged a friend to give him a mean colt he could not sell. Bobby took him on and began to train him. When he mounted him for the first time, the colt threw him off and tried to trample him underfoot. His father had to jump in to save his son. With time and after being thrown off many times, Bobby mastered the colt and sold him at a profit. By age seventeen, Bobby was a robust, self-confident young man.

The question was where to study to launch Bobby's career: Annapolis or West Point? In 1996 a Marine captain and one of Bob's close friends took Bobby to visit Annapolis. He was impressed by its beautiful French Beaux-Arts campus, its history, and its discipline and decided to apply.

Despite its exquisite campus and storied history, however, Bobby found what he saw as the U.S. Naval Academy's (USNA) ambiguous honor concept off-putting. He preferred the West Point cadets to the USNA midshipmen he met on a visit in his high school junior year. His father advised him to apply to West Point and the U.S. Air Force Academy as well as Annapolis. Bobby disagreed. He didn't apply to the Air Force Academy because at that point in his life he wanted to be an infantryman and saw West Point as having an infantry ground–oriented culture.

In the end, West Point won out for three reasons. The first was that because of an eyesight defect Bobby could not become a Marine Corps or Navy flyer at that time. The second was that Bobby saw a real difference between the USMA honor code and that of the USNA. Like some young men of seventeen, Bobby had a very black-and-white concept of right and wrong. To him, West Point's honor code seemed much more clear-cut than Annapolis's more ambiguous concept.

The third reason was because of the personal touch of Susan Baird, one of West Point's Black and Gold civilian representatives. She not only interviewed Bobby personally beside the swimming pool at the U.S. Navy facility near his home where he was serving as a lifeguard but also really encouraged him. Susan was warm, empathetic, caring, supportive, and tenacious. (Her opposite number from the USNA merely telephoned Bobby.) Susan helped Bobby with his application and also arranged for him to participate in an overnight program at West

Point, where he stayed with a cadet. Bobby admired the way cadets carried themselves and interacted with each other. He also admired West Point's culture, academic and physical rigor, and proven track record of developing great leaders including his own heroes, Gen. Dwight David Eisenhower, Gen. George Patton, and Gen. Douglas McArthur.

In the end, Bobby applied to West Point for pragmatic as well as moral reasons. The pragmatic reason was that he wanted the best start possible to a military career. Bobby wrote that "I want to be a career military officer and I want to lead. I want to be the best military officer. I've heard West Point trains good leaders, and figure it could give me the best start possible."[4]

The moral reason Bobby applied to West Point was to make a significant difference in the world: "Right now what I want out of life is to make a difference. This is partly why I want to attend West Point. Certainly, Ike Eisenhower made a difference, as well as George Patton, and also Douglas MacArthur. These men were always my heroes, and maybe one day I will join their ranks in the history books."[5]

The West Point interviewer's notes also reveal that Bobby had not only given careful thought to West Point's renowned honor code but also had already internalized it. When he was asked whether he would turn in a friend and fellow cadet if he caught him cheating, Bobby's reply was emphatic: "I wouldn't have a friend like that. If I did, I would give him a chance to do the right thing, but would turn him in if he didn't. I wouldn't take any pleasure in doing so."

West Point gave Bobby early acceptance (a promise of acceptance subject to his getting a nomination from a U.S. congressman or senator) in the summer between his junior and senior years in high school. The USNA was slow off the mark, only accepting him several months later. Bobby was on his way to West Point.

5

STEPHEN TANGEN

At age eighteen Stephen Tangen was a warm, humble, earnest lad, the second of three sons and the descendant of Norwegian immigrants, a midwesterner to his core. Six feet tall, broad shouldered, and with striking brown eyes, brown hair, and a radiant smile, Stephen was a natural leader: a devout young man of exceptional moral character and unusual moral courage. Modest, he never sought attention much less accolades for his many achievements. No pranks, no adolescent high jinks, Stephen was an Eagle Scout and captain of his high school swimming team, voted most valuable player by his teammates. Swimming was his passion, and he was good, very good. This would prove to be his pathway to West Point.

Born in Hinsdale, Illinois, in March 1986, Stephen grew up in nearby Naperville. In the 1980s and 1990s, Naperville was a vibrant small city with an old-fashioned small-town feel. Located thirty miles west of Chicago, south of Interstate 88, Naperville was a prosperous successful town with average household income of $100,000, excellent public schools and libraries, solid multibedroom homes with manicured lawns, and clean, well-maintained streets. Naperville's schools were a special source of pride. Average scores on the ACT and SAT college entrance examinations were among the highest in not only Illinois but also the United States. Graduation rates too were among the best in the state and the nation. There was also a strong connection to West Point. Three members of Stephen's high school graduating class were accepted to the USMA, two of them in his older brother Andrew's class.

Naperville's downtown buzzed with life: great shopping, trendy boutiques, and over forty restaurants. And it was beautiful. Its architecture was an eclectic mix of styles from elegant Victorian mansions to ultramodern chic. At its heart, there was a beautiful tree-lined river walk on the banks of the DuPage. Above all, people looked out for one another; there was a real sense of community pride. Even though Stephen's family was not as prosperous as most families in Naperville and lived in its unincorporated section, it was an idyllic place to grow up.

Stephen's mother, Susan, the daughter of Illinois farmers, was a dedicated surgical nurse who worked in the Amita Health Adventists Medical Center in Hinsdale for thirty-five years. To ensure that she had enough time to care for her three boys, she worked three days a week. Rising at 4 a.m., she left home at 5 a.m. to prepare the operating theater. Throughout the day she assisted the hospital's orthopedic surgeons, returning home not long after Stephen and his brothers got home from school. She was a loving, devoted mother who ran the household. Susan set a high example of what being a selfless person means.

Stephen's father, Andrew, had had a tough life with several serious setbacks. Born in June 1945 in Twin Creeks, Wisconsin, Andrew was the grandson of Norwegian immigrant farmers. His grandfather was a skilled carpenter who helped build custom staircases for elegant Lake Shore mansions. The first setback occurred during the Vietnam War, when Andrew enlisted in the Air Force rather than wait for the draft. His unscrupulous recruiter told him he could become a pilot. He didn't. Instead, he became an aircraft mechanic servicing hurricane-hunter aircraft in Puerto Rico and had to crawl into the fuel tanks to check and repair them without any facial mask to protect his lungs. As a result, he later developed pulmonary fibrosis. Andrew was proud to have served, but his bitter experience meant he was never an advocate for either of his sons pursuing a military career.

Some years later back in civilian life, Andrew used all of his savings to buy a car dealership, but his business partner betrayed him, and Andrew lost everything. He continued to work in the car industry, successfully managing the finances and sales divisions of other dealerships. To supplement his earnings, he worked as a carpenter in the evenings and on weekends. He left home at 5 a.m. and returned at 11 p.m., six days a week. He saw his young sons only on Sundays when he played with them and told them stories. Throughout Stephen's early life, Andrew was a model of resilience, steadfastness, and selfless devotion to his family.

Andrew and Susan's influence can be seen in the young Stephen's intense religiosity and unusually strong moral character. The Tangen family were active members of the First Congregational Church (United Church of Christ) of Downers Grove, Illinois. When Stephen was old enough he joined the church's youth ministry, where he participated in spiritual growth retreats and youth forums, supported the Christian missionaries, and led fundraising drives for the youth ministry. Stephen went far beyond requirements in the church's two-year confirmation program. He participated in field trips to other Christian churches in the area and completed service projects supporting overseas missionaries and the local community. Later, one of his pastors wrote that "Stephen has grown up before my eyes at First Congregational Church. . . . I have never seen him stray from any of the ideals that are a part of his learning here. People of all ages are fond of Stephen because he seems to care so deeply for us all."[1]

Together, Andrew and Susan set a great example for their three sons not only as selfless parents but also as devout Christians and responsible moral people driven to volunteer in their church and in the wider community.

When Stephen was a boy, it was clear that he was intellectually gifted. He was always observant, ever curious about how things worked and why. He was also the kind of questioning student educators love: the boy who always wanted to learn more than was asked on any given test. Stephen disciplined his mind to become a clear thinker who could break down complex problems into their component parts and then explain the way in which he had done it and why. He was also a selfless, cooperative learner willing to go out of his way to help weaker students in his class succeed.

In Stephen's high school years, what really stands out is that he won so many academic and athletic honors and yet remained so humble. Academically, his teachers ranked him among the top 1 percent of students they had ever taught, awarded him outstanding grades, and applauded his election to the National Honor Society and as an All-American in both the academic and athletic fields in his junior and senior years. In athletics, Stephen received All-Conference awards in cross-country, swimming, and water polo. Yet as one of his teachers wrote, "He never flaunts any honors he receives. His quiet and sincere manner around others belies the amount of recognition he has received in many areas."[2]

But because he stood in the shadow of his older brother, Andrew, Stephen did not yet stand out as an emerging leader. Stephen's leadership drive was nurtured by Boy Scouts and competitive swimming and by two remarkable mentors, Joe LoPresto, his scoutmaster, and Dick Robb, his high school swim coach. A graduate of the University of Chicago's prestigious Booth School of Business, LoPresto had been a highflier at IBM but, disillusioned with the corporate world, retired early to devote himself to community service including the Boy Scout troop based at St. Raphael's Catholic Church in Naperville. There, he gave Stephen and the other boys a master class in leadership by example. As Stephen recalls, "He was very observant, very soft-spoken, very kind, but stern when he had to be. He pushed everyone in the direction they needed to develop. He definitely had that effect on me." LoPresto spotted Stephen's leadership potential when he crossed over from Cub Scouts and encouraged him to stand for election to leadership positions. As early as eighth grade his fellow scouts elected Stephen senior patrol leader. He also served as troop quartermaster, the principal logistician for scout expeditions. Under LoPresto's mentorship, Stephen completed his Eagle Scout project: repainting the Victorian-era bandstand and gazebo in Naperville's historic district. He not only led the repainting but also carefully restenciled and restored the frayed Victorian stars on the bandstand's awning.

Stephen's other mentor was Dick Robb, an eighth-grade science teacher at Jefferson Junior High School and master swim coach at Stephen's high school, Naperville North. There, Robb had turned Naperville North High School into a swimming powerhouse, with state champions in 1995 and 1996. Under his dynamic leadership, Naperville North always had athletes place in the finals and earn All-American accolades. Robb met Stephen at the beginning of his freshman year and spotted his leadership potential immediately. Later Robb recalled that "Stephen was not a verbal leader; he always led by example. He was a quiet influence on the team. He showed great empathy towards other swimmers on the team and handled the clowns with grace. Stephen was always the gentleman."

Stephen's love of swimming began when his mother signed him up for swimming lessons at the local pool. He began competitive swimming at age five. Seeing his interest and aptitude, Susan introduced him to club swimming in the winter. Competitive swimming proved to be another important source of development for Stephen. Like track and field, it's an unusual sport in that it is very individual but also has a team aspect. For Stephen, swim training gave him time alone to think, to reflect on his life and the direction it was taking.

In his freshman and sophomore years in high school, Stephen emerged as the leader of his swim team. Every day he was the first there for practice and the last to leave the pool or the weight-training room. In his senior year, his teammates voted him captain and most valuable player. Perhaps the greatest compliment to Stephen as a leader of the swim team was that he inspired two of his freshmen swimmers to apply for and secure places at the USNA.

In his freshman year, Stephen also played water polo for Naperville North High School. It was here as a freshman that he gave an example of his emerging leadership capability. The school's water polo team lost a game to a rival school's team, one of the best in Illinois. It was a painful loss. Stephen's coach and teammates—all of whom were upper classmen—were so angry by the loss that they refused to congratulate the victors. Stephen too was disappointed but was appalled at the lack of sportsmanship and basic decency displayed by his coach and teammates. He confronted them over their bad behavior and persuaded them to do what was right and congratulate the victors. Later that evening LoPresto, whose son had played for the victorious team, spoke to him about what had happened. Many years later LoPresto recalled, "I realized how strong his leadership skills and values are and that Stephen would go on to be one of America's great leaders."

Will Israel, Stephen's lifelong best friend, was also a straight arrow, the son of conservative middle-class parents from nearby Wheaton, Illinois. They had met through Stephen's older brother, Andrew. They were all part of the local swim club that used the local YMCA pool. Will and Andrew were solid swimmers. Stephen was in the fast lane. He was a high school sophomore, as was Will. They bonded through shared values, a shared commitment to achieving their goals, and a strong moral compass. They supported each other.

Unlike so many other West Pointers, there was no long military tradition to speak of in Stephen's family. But as was the case with almost all young men of the Greatest Generation, his grandmother's four brothers served in World War II, two in the Army and two in the Marines. Stephen's grandfather served in the Navy. Family lore has it that one of them served as Gen. George C. Patton's driver for a short time.

It was Andrew Tangen who planted the idea of military service. At age ten he began reading books about the Navy SEALs and in time decided that he wanted to serve. There was no one decisive moment; Andrew's desire to serve grew steadily over time. At the beginning of Andrew's senior year at Naperville North High School, the 9/11 terrorist attacks occurred, intensifying his desire to serve. Demonstrated passionately by flying an American flag off of the tailgate of his Jeep Wrangler during the months prior to his entry into service. Andrew applied to the USNA but was not accepted. The Citadel, however, not only accepted him but also offered him a Navy ROTC scholarship, and he majored in political science. Slowly as Andrew succeeded at the

Citadel, his father's attitude toward the military began to change. He began to see that after Andrew earned his degree from the Citadel, military service would offer him a meaningful and rewarding career as a naval officer. During his naval career, Andrew served as a surface warfare officer, with special assignments to board vessels and search and seize contraband and in intelligence and counterintelligence operations.

Stephen was now sixteen but had never thought of West Point. Instead, he began to think about Annapolis. Stephen visited Annapolis as a member of his high school's water polo team for a weeklong summer water polo camp. He was captivated by its beautiful late nineteenth-century French Beaux-Arts architecture, its discipline, and its ethos of selfless service to the nation. Stephen and his teammates had a wonderful experience. They lived in the midshipmen's barracks, ate heartily in the midshipmen's mess, and met members of the USNA's water polo team.

Annapolis inspired Stephen to serve. Will Israel, Stephen's best friend on the B. R. Ryall YMCA Swim Team, reinforced his innate desire. Will's older brother, Matt, was a midshipman at the USNA, and Will wanted to follow him there. Will's enthusiasm for the USNA was infectious. And it came at the right time. Now a high school junior, Stephen realized that his family's limited budget meant that he would have to find a way to pay for his own college education. An ROTC scholarship was one option, and Annapolis was another. So, Stephen started the arduous two-year application process to the USNA. It was at this time that he discovered West Point.

Initially thinking of West Point as a second option, after reading Stephen Ambrose's history of West Point, Stephen Tangen began to change his mind. From Ambrose's book, Stephen learned about the USMA's storied history. He was inspired not only by its links to America's Founders, especially Thomas Jefferson's role in the USMA's founding, but also about the enormous contributions its alumni have made to the U.S. Army and to American society as a whole. So, Stephen applied to the USMA as well as Annapolis. He had very good grades in high school but did not have outstanding SAT and ACT scores. This, however, was offset by his strong leadership skills: he had earned his Eagle Scout award in Boy Scouts and was now captain of his high school's swimming team. In October of his senior year, Stephen contacted the swimming programs at West Point and Annapolis. West Point invited him to visit; Annapolis did not—their loss. Swimming had always been Stephen's passion. He had begun competitive swimming at age five and was a highly successful swimmer throughout his boyhood and teenage years. In October 2003, Stephen spent a weekend with the West Point swimming team observing their training and meeting members of the team. During that weekend, there was a minor incident that showed Stephen's drive. As Stephen later recalled, "I was sitting with one of the other kids. There was a bunch of other swimmers that went over to the diving area and laid under the mats pretending to be asleep when the coach was telling everyone to get into the pool. I just got riled up and ran across the pool deck and jumped on them on the mats. Dennis Zilinski, the captain of the West Point swim team, told the coach 'sign him up.' Lo and behold two months later, I got the acceptance letter in the mail."

But Stephen's connection to West Point went beyond swimming. During his visit to West Point, Stephen fell in love with the USMA's culture, academic rigor, capacity to develop outstanding leaders, and above all, moral values. He knew that this was where he belonged, as he stated in his application essay:

> My passion to attend the USMA comes directly from the constant struggle that I undergo to become the best person that I can be. The USMA ... can provide me with a challenging academic, physical, and moral environment, which will instill the virtues of a great leader within me. West Point is also an institution that values and upholds the same morals that I have developed through scouting and my path to become an Eagle Scout. ... Over the last seventeen and a half years I have been seeking for a purpose and direction to follow; that purpose ... of a military and moral leader is exactly what I want to achieve.[3]

Stephen was accepted to Annapolis but with the condition that he attend the USNA's prep school first. After 9/11, with his commitment to West Point's moral values reinforced, Stephen wanted to fight for his country without delay. To do that, he needed service academy education and training. So, it was an easy decision; he was going to West Point.

PART 2

THE U.S. MILITARY ACADEMY

6

WEST POINT

The USMA sits high atop a cliff at an S-shaped bend of the Hudson River. It's a place of striking but austere beauty: a gray granite castle almost medieval in style with parapets, turrets, towers, and dry moats. This is overlooked by the magnificent cadet chapel with the historic lush green grass of the Plain at the center of the campus. At the eastern edge of the Plain is Trophy Point. There on display are captured cannons, heavy mortars, and other military trophies from the American Revolutionary War through World War I. At the heart of the display, towering above the other military trophies, is Battle Monument, a magnificent tall, circular, polished-granite tower honoring the officers and enlisted men of the U.S. Army who died defending the Union during the American Civil War. All around, the views are majestic: to the north the granite Storm King Mountain, to the east the majestic Hudson River and the rugged mountains of the Hudson Highlands, and to the south the densely forested Bear Mountain. This is where our five elite souls were educated and developed into outstanding leaders of character. You need to know something about it.

West Point strives to be the citadel of America's most noble values: duty, honor, and country, yes, but also courage, idealism, integrity, loyalty, modesty, and selflessness. Not every cadet lives up to these ideals, of course, and West Point itself is imperfect. No institution is perfect because every institution is run by people who themselves are flawed. Keep in mind that West Point is not only a military institution but also a topflight college. Like any college or university, it has thousands of attendees ages eighteen to twenty-two years trying to deal with the normal challenges of teenagers growing into adulthood. The vast majority succeed in dealing with these challenges; a minority don't.

West Point has always had challenges. How could it not? Because it admits cadets from every part of the United States, it inevitably imports in each new class the tensions and fissures that exist in American society. West Point's job is and always has been to ensure that whatever the values of entering cadets are, they leave with a set of values found in leaders of character, integrity, honor, and professionalism. This has never been an easy task.

As a result, everything at West Point is all so much harder. When you consider the unique stresses of being a cadet—meeting the exacting academic, military training, and physical fitness standards—you get some idea of how hard it is.

Everything about West Point is rigorous: the admissions standards, the academic curriculum, and the military and physical training. West Point is intense, relentless in its pursuit of intellectual, academic, and military excellence. Cadets' lives are defined by rigorous rules, duty, responsibility, and tradition. West Point is always challenging cadets, giving them little vacation time and filling almost every waking hour with prescribed duties. The job of West Point faculty and staff is to help cadets become outstanding young leaders and officers in service to higher and nobler ideals.

The USMA's core mission is to build leaders of character. In this respect, West Point could not be more different from most civilian U.S. colleges or universities. In his book *The Second Mountain: The Quest for a Moral Life*, David Brooks suggests that in too many civilian colleges and universities today, "students are taught to engage in critical thinking, to doubt, distance, and take things apart, but they are given almost no instruction on how to attach to things, how to admire, to swear loyalty to, to copy and serve. The universities, like the rest of society, are information rich and meaning poor."[1] In contrast, West Point is information and meaning rich. At the USMA, cadets are taught to think critically, holistically, and to "take things apart"; they are taught to be creative and effective problem solvers. But they are also taught how to subordinate their ego, their individual self, to the Army team. Most civilian colleges and universities may not teach "how to attach to things, how to admire, to swear loyalty to, to copy and serve," but West Point does. It teaches cadets how to attach to and admire the USMA's core values; how to swear loyalty to the U.S. Constitution and the values it embodies and how to selflessly serve the nation, its soldiers, and their community. For the most part, West Point succeeds in this. It's no coincidence that seventy-six Medal of Honor recipients are West Point alumni.[2]

The ideal West Point graduate should be a selfless young officer of high moral character, a young man or woman of unquestioned personal honor and integrity. Graduates should be well-educated, tactically proficient, clear critical thinkers able to solve problems speedily and effectively. Above all, a West Point graduate must be able to lead soldiers in combat. Obviously, not all West Point cadets strive to meet these exacting standards, but the vast majority do.

The essence of the West Point development model is that it builds cadets incrementally by raising the pressure, presenting the challenge, and giving them the resources to meet it. The bar is always set at the outer edge of the cadets' capacity to meet it. They struggle mightily to meet it, and when they do, the USMA raises the bar again to a level just beyond their current capacity, once again repeating the cycle of pressure, challenge, and provided resources, and the cycle continues. But for cadets it's a long, arduous four-year process.

In other words, West Point's mission is not only academic and military but also psychological. The USMA takes outstanding young men and women, breaks down their individual egos, and rebuilds them, making them think less as an individual and more as their role within the Army team. The USMA does not destroy their individuality but rather integrates it into the team. West Point plants the idea of the pre-eminence of the team over the individual during Cadet Basic Training, known as Beast Barracks. The USMA then develops this idea through its Cadet Leader Development System. The focus on the team over the individual is relentless and continues irrespective of changes in the cadet leadership cadre. Through this transformation, West Point builds the foundation of the cadet's moral character, the core values that unite the military team: duty, honor, integrity, loyalty, respect, selflessness, and personal courage. The stronger the moral foundation that cadets bring to West Point, the greater the chance that West Point's transformational process will elevate them to be elite souls. Those cadets with a weaker moral foundation greatly benefit from the USMA's honor code. The vast majority of cadets become good officers, and a few become elite souls. One of those elite souls, Ross Pixler, West Point class of 2005, explains the USMA's transformation process with special insight. "The transformation is not quick, but rather a slow, grueling process with highs and lows based on shared experiences and hardship. It is these difficult times which build and strengthen the character and create resilience. Resilience is not created through a PowerPoint or through a classroom training event. Resilience is built upon perspective. The more an individual overcomes or the more difficult a task he or she endures, the greater the impact it has. This is as true psychologically and spiritually as it is physically."[3]

West Point does provide limited recovery periods when cadets can reflect upon and draw strength from their successes. Then it is on to the next challenge, intellectual or physical, and the next hardship. For West Point cadets, it is the beginning of a lifelong process of character building.

Why does West Point demand such a strenuous four-year program? Once again, Ross Pixler provides a key insight. "No one changes when they are comfortable. We must make our future leaders uncomfortable, exhausted, weak, hungry, cold, and wet if we want them to someday be tough, alert, morally straight, and uncompromising in the face of insur-mountable challenges."[4]

I admire West Point, flaws and all. I first visited the USMA in 1985 as a British diplomat leading a Eurogroup team for discussions on NATO burden sharing. I was struck by not only its austere physical beauty but also its nobility of purpose and culture of excellence. As a Briton, I was amazed at how West Point combined the features of a top-flight university with those of a world-class military academy. The intellectual firepower of the Social Science Department faculty I worked with on that Eurogroup visit blew me away. Over the years as I have been a Thomas Hawkins Johnson Visiting Professor and a regular visiting speaker and contributor to Europe's student and senior conferences and its scholarship program and, since 1998, as an adjunct professor, my admiration has only grown. I have rarely met a group of people who love their profession more than the officers at West Point. And I have rarely encountered such self-discipline coupled with such academic rigor and creative policy thinking. There is generally a palpable sense there that everyone is doing something special, and they usually are. In addition to teaching and mentoring cadets and conducting scholarly research, so many West Point officers and alumni have made important contributions to U.S. national security policy in the White House; on the National Security Council; in the Pentagon, the State Department, and the intelligence community; and as generals on the Joint Chiefs of Staff. It's an extraordinary record of achievement.

Among most cadets too I have usually sensed something special. West Point tries to recruit men and women of strong moral character, demonstrated leadership potential, and high academic and athletic achievement. And since the terrorist attacks of 9/11, this meant young people willing to become Army officers in time of war knowing that within a year or so of graduation they would be deployed in harm's way. That takes courage.

Each incoming class is full of high school valedictorians; varsity high school team captains in swimming, football, soccer, cross-country, and track and field; Eagle Scouts; Girl Scout Gold Award recipients; and National Merit Scholars. West Point cadets are a remarkably accomplished group of young people. Among them, I have always found a cheerful can-do spirit: young people who take justified pride in their mental and physical toughness and their capacity to manage a level of stress that most civilian students could never imagine, let alone cope with. Author David Lipsky, who spent four years at West Point research-ing his book *Absolutely American: Four Years at West Point*, got it exactly right on cadet values: "You hear cadets talk about 'honor,' 'character' 'achieving excellence,' 'selfless service,' 'principles,' 'developing yourself,' and 'leadership' without a flicker of a smirk."[5]

The cadets, however, are not perfect. Not all of them are highly motivated strivers; some have mastered the unfortunate art of doing just enough to get by, while others willfully break the rules. In 2020, seventy-three cadets cheated on a mathematics final. Failures to meet the standards of personal and ethical conduct expected of cadets by the institution is not surprising; perhaps what is surprising is the rela-tive infrequency with which it happens. That there are problems now is as true as it was when the five elite souls were cadets at West Point between 2001 and 2008.

Back then, however, there were three particular problems. The first was friction over the presence of women in the corps. The first classes of women at West Point in the late 1970s and 1980s had suffered an appalling level of discrimination and hazing from their male coun-terparts in the corps. Between 2001 and 2008, despite vigorous efforts by successive superintendents and military and civilian faculty, there were still some misogynistic male faculty and cadets who believed that women had no place at West Point.

The second problem when the elite souls were cadets was sexual harassment and assault of female cadets. With over 4,000 vigorous young men and women clustered together, normal consensual rela-tionships were always going to happen and did. These were not the problem. The problem for West Point was that like any college or uni-versity, it imported wider societal problems along with the 1,200 new cadets from every conceivable background. Teaching the young men to treat their female fellow cadets and subordinates with respect was (and remains) a constant challenge.

The third problem when the five elite souls were cadets was alcohol abuse. Until the late twentieth century, West Point strictly prohibited alcohol consumption. But as American society's attitude toward alcohol abuse began to shift away from zero tolerance, so did the Army's and West Point's. Brig. Gen. Dan Kaufman (Ret.), West Point's former dean, summed up the problem:

> Cadets who graduated during the "no alcohol at all" era predictably responded to the "forbidden fruit" syndrome by overindulging when they got to the real Army. As the Army's attitude changed, lots of young officers had a blemish on their record right off the bat. West Point, wisely in my judgment, decided to tackle the "forbidden fruit" syndrome by allowing cadets of legal age to consume on post at certain times. I used to joke when I was Dean that during the first month of school each year a good deal of the senior class would end up in my front yard as they struggled back from their club after overindulging in their newfound freedom. By the end of that first month, however, the number essentially disappeared. The cadets learned that consuming alcohol costs money, makes you feel bad, and gets you behind in your schoolwork. So, they figure it out in an environment where making a mistake is not catastrophic. The results of the program were significant: the incidence of DUI and other alcohol-related offenses by new graduates once they reached the Army dropped dramatically. The problem for West Point is, of course, that every year it has to teach the same lessons to a new crop of young adults. Do some cadets suffer from alcohol abuse? Certainly, but the cases are episodic, not chronic.[6]

Two of the five elite souls believed, however, that when they were cadets in the years after 9/11, there were two different standards and that preferential treatment was given to cadets who played football and other collegiate sports. One of the elite souls even suggested that alcohol abuse was "a huge problem." He alleged that there was no coherent policy, no equal standard of justice. Football players got away with drunkenness while other cadets were severely punished. The other three elite souls said they never witnessed this.

The Honor Code and the Honor System

At the very heart of West Point's culture is its storied honor code, which states that "no cadet shall lie, cheat, steal, or tolerate those who do." In the early twenty-first century, West Point's honor code and reformed honor system drew the five elite souls like a magnet attracts metal. By age eighteen, all five elite souls were not only accomplished young leaders but also young men with an exceptionally strong moral foundation. All five had strong moral reasons for wanting to attend West Point and had given careful thought to the USMA's honor code and honor system. To a large degree, all five had already internalized it. Ross Pixler, for example, thought of the USMA in almost religious terms, seeing it as the way for him to answer his "calling." Stephen Tangen wrote in his application that his passion to attend West Point came from the fact that his honor code and West Point's were identical. Bobby Sickler respected West Point's clear-cut honor code. Nick Eslinger admired the moral character of the West Point cadets who visited his high school. And Tony Fuscellaro highlighted the compatibility of his moral values with those of West Point. The five elite souls benefited from reforms in the implementation of the honor code.

The USMA's founding father, Col. Sylvanus Thayer (superintendent, 1817–1833) believed that West Point's most important goal was building U.S. Army officers with character. To that end, Thayer believed that cadets must have self-discipline and moral uprightness. To build self-discipline, a cadet had to successfully complete West Point's rigorous academic, military, and physical training programs. To build moral uprightness, cadets had to strictly adhere to the honor code. Other cadets and an honor committee monitored cadet compliance. Anything less than 100 percent compliance meant expulsion.[7] Unfortunately, this harsh approach was not matched by teaching cadets how to interpret and apply the honor code to the complex morally ambiguous situations that arise in life.

But did all cadets really absorb West Point's honor code? The majority did absorb and internalize the honor code. They became true believers. Others, however, merely paid the honor code lip service or abided by it, lest they be punished.

After the 1976 cheating scandal, the turning point in West Point's approach to its honor code came in the conclusions of the Borman Commission report published in early 1977. West Point alumnus and Apollo astronaut Col. Frank Borman (Ret.) and his fellow commissioners criticized the inadequacy of honor education at West Point as well as the unfairness and even corruption within the honors committee itself.[8] They recommended that the superintendent be given flexibility in deciding the most appropriate punishment for cadets who violated the USMA's honor code.[9] That same year, Army chief of staff Gen. Bernard Rogers launched a root-and-branch review of every aspect of West Point including the honor code. Together the three committees that he established became known as the West Point Study Group.[10]

In terms of personal behavior, the Borman Commission and the West Point Study Group recommended that the existing honor code be the beginning, not the end, of a process of character development. As a result, West Point's leaders concluded that better and more flexible enforcement of the existing honor code had to be supplemented with an honor education program that would help cadets think through appropriate ethical responses to complex moral problems that had no simple right or wrong solution.[11] Consequently, the USMA launched a series of important initiatives that transformed character development and honor education at the USMA. At the heart of these reforms was the idea of incorporating lessons in ethical decision making into class syllabi across all disciplines.[12]

Lt. Gen. Dave Palmer (superintendent, 1986–1991) took the next step toward reforming the honor system. He integrated the building of ethical cadets into his new Cadet Leader Development System by requiring every department at the USMA to share the responsibility for cadets' development as honorable officers.[13] Two years into his tenure as superintendent, General Palmer asked then U.S. Army chief of staff, Gen. Carl Vuono, to establish an external commission to provide an outside evaluation of the recommendations of the internal committees that had examined the honor code. In particular, General Palmer wanted the outside commission to evaluate the effectiveness of the initiatives to integrate ethics into all programs at the USMA, which was a key goal of his Cadet Leader Development System.[14] Chaired by Wesley Posvar (West Point class of 1946), then president of the University of Pittsburgh, the commission delivered a generally positive assessment of the USMA's honor system and character-building program.[15] One recommendation of special importance to one of the elite souls was that the superintendent should have even greater discretionary authority in dealing with cadet infractions of the honor code. If the superintendent had not had that authority, one of the elite souls would have been expelled from the USMA.

General Palmer's successors as superintendent, Gen. Howard Graves (1991–1996) and Gen. Dan Christman (1996–2001), further strengthened the USMA's honor system. General Graves instituted the honor and respect programs and brought in additional staff to oversee and teach both.[16] By 1995, these two programs provided a total of 113 hours of instruction in ethics and respect for others.[17] In 1998, three years before the first three elite souls arrived for Reception Day (R-Day), Lt. Gen. Dan Christman created the Center for the Professional Military Ethic (later the William E. Simon Center for the Professional Military Ethic). Well staffed and well resourced, the new center's mission was to integrate instruction for honor and respect across the curriculum.[18] General Christman also introduced the Honor Mentorship Program, a six-month program in which a cadet found guilty of a lesser honor code violation was required to study a prescribed course of readings about what honor really means. In the spirit of self-analysis and rehabilitation, the cadet then discusses the readings in a weekly tutorial with a military mentor.[19]

By the early twenty-first century, when the five elite souls entered West Point, the harsh way of building honorable officers had given way to a more educational and flexible approach. The five elite souls had all the ingredients to be great officers: outstanding intelligence, strong character, physical and moral courage, physical fitness, and a passionate desire to excel. West Point's reformed honor system helped them build on their strong ethical foundations.

Getting that Coveted Place: Admissions

Getting into West Point is very difficult. On average, nine out of every ten applicants fail. You begin in your junior year in high school by completing a candidate questionnaire so that a USMA admissions officer can decide whether or not you are qualified to compete for a place at the USMA. Once you have completed your application dossier, you have to get a nomination by your U.S. congressman, one of your state's two U.S. senators, or the vice president. There are also a small number of service-connected nominations: if you are an ordinary soldier on active service, you can be nominated by your platoon or company commander. You can also receive a service-connected nomination if your parents are on active duty or if one of them received the Medal of Honor or was completely disabled as a result of combat wounds. Candidates must also complete a medical exam, a rigorous physical fitness test, and the SAT or ACT with writing.

Previously, congressmen and senators could rank order their nominations, and if their top nominees were fully qualified, the USMA had to admit them. At the same time, West Point had the opportunity to select qualified alternates from the congressmen's or senators' list. Over time many congressmen and senators stopped ranking their nominees, thereby giving the USMA more freedom to select the most qualified nominees. In the late 1960s when Congress expanded the number of cadets that West Point could enroll from 2,500 to 4,000, this gave the USMA even greater freedom to choose the best-qualified alternates.

By the early twenty-first century when the five elite souls applied to West Point, the USMA had introduced waves of reform designed to recruit the best possible candidates. As former West Point historian Lance Betros has written in his excellent work *Carved from Granite: West Point since 1902*, "The greatest changes in the admission system took place after World War II. Academy leaders worked with Congress to improve nomination procedures and thus attract higher-quality candidates than before. Concurrently, they devised better methods to evaluate the character, intellect, physical fitness, and leadership abilities of the candidates. The Academy underwent organizational changes to strengthen recruitment, expand publicity, and improve administrative support."[20]

In reforming the USMA's entry requirements, perhaps no one was more important or more relevant to the elite souls than Lt. Gen. Garrison H. Davidson, superintendent from 1956 to 1960. It was he who developed and implemented what became known as "the whole man system," which has remained in effect ever since.[21] The system set balanced entry requirements that embraced intellectual ability, physical fitness, and leadership. This greatly smoothed the admissions path for the five elite souls, all of whom embodied the three qualities desired by General Davidson.

In 1971 another energetic reformer, Director of Admissions Manley Rogers, launched a recruitment initiative without which Ross Pixler, one of the five elite souls, would not have been admitted to West Point. Rogers supplemented the District Representative Program (staffed by alumni volunteers) with the Military Academy Liaison Officer (MALO) program. These military liaison officers were Army reservists, mainly West Point alumni or former faculty members; their principal task was community outreach and recruitment.[22] As we saw in chapter 3, one of these MALOs, Lt. Col. Jack Ruffey (Ret.), rescued Ross Pixler's application when the USMA had declined to admit him in the mistaken belief that he had asthma.

Another of Rogers's successful initiatives helped persuade Nick Eslinger, another elite soul, to consider attending West Point. Understanding the importance of peer-to-peer influence, Rogers significantly expanded the Cadet Public Relations Council.[23] So, by the 1980s over one thousand cadets fanned out every year across the United States to speak at schools, Boy Scout Jamborees, student conferences, and teacher conferences.[24] As Nick told author J. Pepper Bryars in an interview for his book *American Warfighter*, two cadets from the Cadet Public Relations Council visited Nick's high school in Mountain View, California. To him, they looked the epitome of successful young men who described West Point as "a leadership factory." This argument that West Point was an outstanding center for leadership development really captured his attention. Nick's mind was made up.[25]

But it was another of West Point's post–World War II reforms that helped Nick get there: the USMAPS founded by the superintendent, Gen. Maxwell Taylor, in 1946.[26] Nick had everything West Point wanted: good academic grades, great moral values, real athletic prowess, and an outstanding leadership record. But he struggled with the SAT exam. So, West Point offered him a place at the USMAPS, then located at Fort Monmouth, New Jersey. The goal was to help him improve his academic performance to the standard required by West Point. The USMAPS did that and more. This is how Nick, the excellent student, outstanding young leader, and varsity athlete who did not test well, made it to West Point.

Analysis of the application dossiers of the five elite souls in the USMA Archives shows that at the beginning of the twenty-first century, West Point had several key admissions criteria: outstanding academic ability, leadership in extracurricular activities other than athletic, excellent athletic ability, and personal qualities as well as strong character.

Specifically, in terms of academics, the most important single admissions criterion the USMA admissions team was looking for was young men and women at or near the top of their high school class (the more academically rigorous the high school the better), those who were critical thinkers and sought out intellectual challenges beyond those required in their coursework and had earned high academic honors such as a National Merit Scholarship.

Turning to leadership, West Point wanted those who had been creative and original leaders who had made a real contribution to their community. In athletics, West Point was seeking candidates who at least earned a varsity letter in a sport or were excellent prospects for one of its varsity athletic teams.

Finally, turning to personal qualities, West Point sought candidates with a positive, outgoing personality who were well-rounded, well-spoken, and congenial. West Point also wanted candidates who were empathetic, collaborative team players, effective communicators (verbally or in writing), and of high moral character who would always take responsibility for their actions, were able to perform well under pressure, and could handle a demanding schedule without neglecting their academic work.

In different ways, the five elite souls met these criteria. Academically, they were all very capable, talented young men, although one struggled with the SAT. None, however, had won national academic honors, although one showed great intellectual promise despite the limitations of his underresourced West Virginia high school. What they all had in common was great intellectual curiosity, a drive to seek out new intellectual challenges beyond the requirements of their regular coursework. All five were good athletes, and two were exceptional. They had earned varsity letters and captained their high school teams; one was an All-American scholar-athlete. All five demonstrated great leadership potential and already had impressive track records as serious contributors to their communities. Above all, every one of the five was a young man of exceptional moral character.

R-Day

The cadet career at West Point begins on R-Day in late June just a couple of weeks after high school graduation. When the five elite souls entered (2001, 2003, and 2004) and today, R-Day is a shock: jarring and disorientating. In a series of steps over a five-hour period, R-Day transforms eighteen-year-old civilians into cadet candidates. Physically, these high school valedictorians, class presidents, Eagle Scouts, and champion athletes are stripped of their individuality: their clothes, their personal possessions, and, for the men, even their hair.

Psychologically, the shock is just as great. It begins when cadet candidates are given a mere sixty seconds to say goodbye to their parents. Then these hometown stars, once lionized by their peers and their community, are psychologically challenged. Escorted by cadets, they proceed from station to station—medical screenings, barbershop, drawing Army-issue clothing—wearing USMA Form 2-176 around their neck and being verbally challenged by the cadet cadre in charge of R-Day. The new incoming class is made to feel like a five-year-old again on their first day at elementary school. The USMA form is the size of a note card and has boxes to be checked off by the cadet cadre as each new cadet candidate completes tasks at each station. These high school stars each began R-Day as a treasured child, a respected future leader, an admired member of their local community. They end the day absorbing the unpleasant truth that they are no longer exceptional, no longer in charge of their life or working environment. Instead, the message from West Point is that you are no longer just an individual, you are part of a larger whole: the USMA Corps of Cadets and the United States Army. And if you think you are special, prove it.

The purpose of this difficult day is to break down cadet candidates' individuality. But it's also about building them up. As *New York Times* columnist David Brooks writes in his classic book *The Road to Character*, moral heroes have "to descend into the valley of humility to climb to the heights of character."[27] This is what R-Day is really about: it begins the process of building character by a sharp shock designed to help new cadets get rid of their adolescent ego and self-centeredness and take full responsibility for all that they do, no matter the consequences.

Beast Barracks

Before the new cadet candidates can recover from the physical and psychological shock of R-Day, they begin Cadet Basic Training. More popularly known as Beast Barracks, it's just over six weeks of physical and mental purgatory to see if the new cadet candidates have the strength of body and mind to cope with exhaustion, sleep deprivation, stress, torrential rain, and high humidity and to learn new military skills. In his book *Duty Firs : A Year in the Life of West Point and the Making of American Leaders*, former West Point professor Ed Ruggero describes Beast Barracks as "six and a half weeks to learn how to look, walk and talk like soldiers; to begin to absorb—or be absorbed by—the military culture; to learn soldier skills, anything from how to march to how to fire a weapon; to learn how to obey."[28]

Beast Barracks is divided into two equal parts. The first three weeks take place largely on West Point's gray granite campus. For the cadet candidates it's about daily physical training (push-ups, sit-ups, jumping jacks, timed runs around West Point's hilly campus), road marches up to six miles, team building, and learning the Army's culture, the USMA's unique traditions, and to obey orders. There are so many things to learn and memorize: the "knowledge" book, how to eat in the USMA's cathedral-like mess hall, how to salute, how to organize and clean your room, and how to dress.

The second part of Beast Barracks is all about basic Army field training and takes place at Camp Buckner from mid-July through early August, when the Hudson Valley's weather grows ever hotter and more humid. In his book *The Unforgiving Minute: A Soldier's Education*, West Point alumnus Craig Mullaney wittily describes Buckner as "summer camp with automatic weapons."[29] Army field training at Bucker is that and more. It's a grueling test for youngsters less than two months out of high school and involves mountain climbing, rappelling down a fifty-foot cliff, and coping with being tear-gassed. Beast Barracks II also involves a lot of basic infantry skills: learning how to fire a rifle, capture a hill from the enemy, capture a trench, provide covering fire for your fellow soldiers, and move around safely under enemy fire and participating in nighttime patrols under simulated fire from Army regulars. Beast Barracks II finishes with a fifteen-mile march from Camp Buckner back to West Point.

Beast Barracks is not only about testing the minds and bodies of the cadet candidates but is also about developing leadership in the upperclassmen. For them, the question is whether or not they have the instinct and the skills to lead effectively. If they are beginning Yearlings (second-year cadets), can they lead a squad of twelve? If they are beginning Cows (third-year cadets), can they lead a platoon of forty? If they are beginning Firsties (fourth-year cadets), can they lead a company of about 160? At Beast Barracks, the upperclassmen are carefully scrutinized to see if they are leading by example, to see if they are servant leaders who put the needs of their cadet candidates ahead of their own. If the cadet candidates under your command succeed, so do you. If they don't, you will be sharply criticized by the Army training, advising, and counseling (TAC) officers supervising Beast Barracks.

Cadet Leadership Development Model

When the five elite souls arrived at West Point (three in 2001, one in 2003, and one in 2004), they benefited from a leadership development model that had been finely honed over several decades. This was the result of many important reforms launched by six West Point superintendents: Gen. Douglas McArthur, Gen. Maxwell Taylor, Gen. Garrison Davidson, Gen. Andrew Goodpaster, Gen. Dave Palmer, and Gen. Dan Christman. Their reforms had produced a holistic approach for developing young tactical officers that was based on the insights of the behavioral sciences and was more attuned to the needs of the Army in the early twenty-first century.[30] Former West Point historian Col. Lance Betros (Ret.) ably summarized the results: "While these initiatives were neither perfect nor a panacea, they allowed West Point to become a true leader *development* institution, as opposed to an unforgiving crucible in which to prove oneself fit for leadership."[31]

The Cadet Leadership Development System

Gen. Dave Palmer, a distinguished soldier-scholar, took command at West Point in 1986. One of the most important problems he faced was that the old Fourth-Class System tolerated a demeaning cadet culture:

hazing and verbal abuse of Plebes and other underclassmen. This was never USMA policy, but everyone knew it went on out of sight of the TAC officers, and some military faculty turned a blind eye. Hazing and verbal abuse were supposed to test the Plebes' mental and physical toughness, to teach them military discipline as well as respect for the military system, but it often got out of control.

USMA leaders were aware of the problem and tackled it frankly in their "Institutional Self-Study, 1988–1989: Report to the Commission on Higher Education Middle States Association of Colleges and Schools in Preparation for the 1989 Decennial Reaccreditation."[32] As Col. Jim Golden, chair of the USMA's Steering Committee, later put it, "The Fourth-Class System designed to put pressure on Plebes was often abused in practice. Just as importantly it taught upperclassmen leadership techniques that were not appropriate in the field army. The recommendation was to convert the Fourth-Class System with its high hurdles in the first year into a Sequential Leader Development System with increasing standards and responsibility."[33]

The goal of General Palmer's reform was to introduce greater professionalism into the relationship between upperclassmen and Plebes.[34] After the reforms in General Palmer's tenure, some faculty critics of the old system thought the Palmer reforms had not gone far enough and that the upperclassmen had not fully bought into them and were only implementing them because they did not want to be punished.

"The Changes"

To a degree, Lt. Gen. Palmer's critics were right to say that his admirable reforms did not go far enough. More were needed. By the late 1990s, there was still abuse and intimidation of Plebes. Many upperclassmen were still not learning how to motivate and inspire soldiers. In 1997 the new commandant, Brig. Gen. John Abizaid, delivered a plan that quickly became known as "The Changes."[35] The new superintendent, Lt. Gen. Dan Christman, fully supported him. The goal of Abizaid's plan was to make the West Point cadet experience more realistic, more like that of soldiers, NCOs, and officers in the field army. Brigadier General Abizaid redesigned the military curriculum and toughened up military training to help cadets better grasp and be prepared for the harsh realities of combat. He wanted West Point to be "demanding not demeaning," and he meant it.[36] Henceforth, there would be greater emphasis on reason and persuasion, a new concept of the commander as inspirational leader.

For the cadet cadre, "The Changes" built on General Palmer's reforms by giving each Yearling, Cow, and Firstie the opportunity to lead by example and inspiration: to learn how to take care of Plebes and how to mentor them and other underclassmen so they could solve their own problems and meet their academic, military, and physical fitness requirements. As former West Point professor Ed Ruggero concludes, instead of harassing or hazing Plebes, now the upperclassmen "must design training that will instruct and motivate. This is exactly what NCO's and junior officers do every day in the Army."[37]

Ruggero also states that "the culture changed from one in which Plebes were held to high standards while everyone else slacked off to one in which the upper three classes are held to increasingly high standards as they progress through the four years."[38] This was a far cry from the West Point of the past. All five elite souls greatly benefited from this new model.

Military Training at West Point

The five elite souls began their military training at West Point after 9/11 in a time of war. This gave them as well as their TAC officers and military science instructors a new sense of urgency. The five young cadets knew they were preparing to do something bigger and more important than the normal peacetime service. Consequently, they each approached their military training with fierce determination, as did the military science division. The commandant and his team worked closely with

United States Army Training and Doctrine Command to absorb the lessons of Afghanistan and Iraq and incorporate them into the military science curriculum: the new doctrines, the new ways of organizing to fight, and the new ways of waging war. To teach these lessons, the commandant brought back young combat veterans who taught the elite souls military science informed by the hard-earned lessons of war. But because of the demands of the two wars in Iraq and Afghanistan, it was challenging for the USMA to keep these combat veterans.

In addition to learning the hard lessons of war, the five elite souls grew and developed through West Point's regular training. They learned to follow before they could lead. They learned to enjoy being pushed to the limits of their physical and mental endurance. From excellent TAC officers and NCOs, they learned to inspire and motivate soldiers. From officers and NCOs in the regular Army, they learned to understand and motivate the soldiers of the all-volunteer force. By the time the five elite souls graduated from West Point, they had some of the Army's best military training.[39]

For most of its existence there have been two different visions of West Point, and the adherents of these two visions have traditionally been called the Spartans and the Athenians. The Spartans believe that the USMA's primary emphasis should be on military training. They argue that West Point graduates should be tactically proficient enough to take up their duties in whatever branch of the Army they have chosen almost immediately after graduation. In contrast, the Athenians argue that West Point's primary purpose is intellectual and character development and that branch-specific professional military training should take place after graduation. The Spartans versus Athenians debate takes on its own intensity not only because cadet time is limited but also because of the demands of war since the turn of the century have given its resolution greater urgency.

Although the training at West Point is not branch-specific, it is more realistic and more relevant to the combat situations that twenty-first-century graduates are likely to face.[40] The five elite souls thought so. They all believe that the USMA's military training equipped them well for their subsequent branch-specific training and for their commands in Iraq and Afghanistan.

When the five elite souls arrived at West Point early in the twenty-first century, they benefited from decades of improvements and innovations dating back to World War II.[41] The driving force behind these innovations was war: the need to adapt cadet military training to equip young USMA graduates with the tactical skills necessary to meet the ever-changing demands of modern war. Among the most important of those improvements and innovations were the following:

- The USMA acquired and developed its own modern military training complex at Camp Buckner, with cadet barracks, firing ranges for a variety of weapons, and amphibious training sites.[42]
- The USMA introduced Cadet Troop Leader Training (CTLT), a program that places cadets about to enter their final year as assistant platoon officers in the field army for three weeks. Although the quality of their CTLT experience varied, the program did give the elite souls the opportunity to serve alongside experienced junior officers and also enabled them to learn more about non-commissioned officers and soldiers.[43]
- Active-duty units became involved in cadet summer training. For the elite souls, this was a transformative experience. They were not only inspired by the effectiveness of the visiting officers and their close relationships of trust with their NCOs but also gained a deeper understanding of the culture of NCOs and enlisted soldiers.[44]
- Cadets about to enter their final year could attend a variety of Army specialty schools.[45] For the five elite souls, shaped by the 9/11 terrorist attacks, any opportunity to gain additional military skills was eagerly welcomed.
- The USMA gave increased opportunities to incoming Firsties to serve as cadet cadre leaders for Beast Barracks as well for cadet field training.[46] All of the elite souls found this to be a great developmental opportunity.

- In 2003, superintendent Lt. Gen. William Lennox brought military science courses back into the academic year so that cadets could have more year-round exposure to military matters and be better prepared for their summer leadership duties.[47] Although this decision was controversial, the five elite souls believe the addition to the curriculum gave them better preparation for their postgraduation military training.

The Academic Program

The wars in Iraq and Afghanistan presented new challenges in the early twenty-first century and placed many complex intellectual demands on Army officers. Writing in the journal *Liberal Education*, West Point professor Bruce Keith explained these new challenges: "Today's military operates in contexts where uncertainty and ambiguity are commonplace. Human security challenges exacerbated within regional trouble spots are frequently characterized by extreme poverty, a lack of infrastructure, an inability of people within these regions to plug into a globally connected world, and intrastate violence."[48]

To meet these challenges, the U.S. Army needed highly educated officers with nimble minds, able to think critically and holistically across disciplines and to effectively solve problems. West Point's reformed academic curriculum gave the five elite souls those skills and more. The curriculum made them better officers because it forced each of the five to study harder, longer, and more widely than ever before; stretched and tested their minds as never before; and introduced them to challenging new academic disciplines. The demands that the curriculum placed on these five young men gave them increased concentration, a capacity for deeper thought, sharper critical thinking and better problem-solving skills. Their engineering courses in particular helped them dig down to the core of a problem, break it down to its smallest parts and see the interconnections between parts but also separate the relevant from the irrelevant.

Two of the five took advantage of new and especially challenging majors: nuclear engineering and life sciences. Three of the five struggled just to be average. Only two excelled academically, but by their fourth year all of them made the dean's list, a testament to their determination.

The five elite souls carved their own distinctive path through West Point, however. Unlike the majority of their classmates, only one of the elite souls chose to major in the humanities, one of the most popular majors, along with the social sciences, at the time. Three of the five chose to major in science and engineering, and one majored in military science. When they had the freedom to choose, the five elite souls chose to emphasize military rather than academic summer options. None studied abroad. After 9/11, these five remarkable young men focused with unwavering determination and intensity on becoming the best possible platoon officers they could be. West Point's academic program was well placed to teach and develop these skills.

When the five elite souls entered the USMA, they found an academic program that was better, more rigorous, and more carefully thought out than ever before, with a clear definition of what a West Point graduate should know and a learning model that was more demanding and more integrated and offered more choice than at any time in the USMA's history.[49] In the academic program as in military training, the five elite souls benefited from many years of sound reforms.

The first major reform was the gradual customization of the academic program that began in the early 1960s.[50] The core curriculum would provide the foundation for cadets' intellectual and professional development, while electives would provide the opportunity to deepen their knowledge and sharpen their analytical skills in areas of special academic interest to them.[51] The addition of electives triggered a vigorous debate over the next twenty years between reformers and traditionalists. But by 1970, electives had proved so popular that the Academic Board allowed cadets to take up to six general electives or cluster them in five areas of concentration: the physical sciences, engineering, the humanities, national security studies, and management studies.

The massive cadet cheating scandal of 1976 brought two major reviews of the USMA's academic program. The analysis and recommendations that came from these reviews had far-reaching consequences for West Point and our five elite souls. The two commissions, the Borman Commission and the internal West Point Study Group, offered a number of conclusions and recommendations. The most important included the following:

- Cadets were overscheduled and did not have enough time to study, much less reflect on what they had studied.[52]
- To create more time for study, military training should be confined to the summer, and the number of courses needed for graduation should be reduced from forty-eight to forty-two.[53]
- Cadets' normal semester academic course load should be cut from six to five.[54]

In practical terms, cutting the number of courses meant cutting the number of required STEM courses and abandoning the USMA's long-held belief that mathematics was the best academic discipline for developing effective problem-solving skills. Feedback from commanders in the field army suggested that West Point's emphasis on mathematics left graduates intellectually inflexible and inclined to search for neat, orderly solutions to problems where they didn't exist.[55] This was not an indictment of the quality of math teaching at the USMA; rather, it was a result of the fact that a lot of cadets did not have a natural talent for math. As a result, they were not able to see the discipline as an act of discovery through critical thinking and logical leaps; they survived by memorizing formulae, and rote memorization is the antithesis of critical thinking.

After intense debate over the number of courses needed to graduate and over the balance of the curriculum between quantitative and nonquantitative classes, the USMA's Academic Board decided to adopt academic majors. With support from the Army, West Point implemented the academic majors in the 1983–1984 academic year.[56]

This model remained intact when the five elite souls began to arrive at West Point.

Another round of reforms that shaped the elite souls' educational experience at West Point were driven by two developments in the late 1980s: a strategic review launched by Lt. Gen. Dave Palmer and West Point's Middle States reaccreditation process. General Palmer's review had three main conclusions:

- The cadets' academic load was too heavy and did not give them enough time to reflect and take ownership of their education. Consequently, General Palmer ordered that the minimum course load to graduate be reduced to forty.[57]
- Cadets majoring in non-STEM disciplines should take more classes in engineering. To that end, cadets would take a five-course sequence in any field of engineering so they would not only learn how engineers think but also apply that insight and knowledge to help them solve the kind of practical problems they would face as officers in the field army.[58]
- General Palmer implemented a strategic planning process to examine two basic questions: What did West Point want its graduates to know? And how well were they learning it?[59]

Ten years after General Palmer's initiatives, General Dan Christman, superintendent from 1996 to 2001, introduced two further reforms dictated by the new needs of the twenty-first century. First, he replaced two of the required engineering courses in the core curriculum with a foreign culture course and an information technology course. Second, he introduced a new interdisciplinary capstone course that was designed to foster a more holistic approach to learning and problem-solving.[60]

Another development that greatly enriched the cadet academic experience was the introduction of "Margin of Excellence" academic programs, the most important of which was arguably the Academic Individual Advanced Development (AIAD) program, designed to help cadets apply concepts learned in the classroom to the real world. Funded by generous private philanthropy, the AIAD program gave cadets fully funded internships during spring break or over the summer. These internships could be taken up in any branch or agency of the federal government, think thanks and other research institutes, and scientific laboratories.[61] Oddly, only one of the elite souls, Ross Pixler, took advantage of AIAD and spent three weeks with the National Aeronautics and Space Administration (NASA) in Houston.

In Retrospect

Looking back, the five elite souls saw their West Point experience as transformative but imperfect. It was a slow, grueling process that demanded all of their determination and resilience to overcome the inevitable frictions and setbacks. We will look at their individual experiences at West Point in the next five chapters. Today the elite souls admire and respect West Point, but some of them believe that when they were cadets, the USMA leaders should have been more open to criticism and should have shown a greater capacity for self-criticism and self-diagnosis. Ross Pixler still believes in West Point. "We can point at splinters in the granite of West Point, but miss the logs present everywhere outside these hallowed halls."[62]

Tony Fuscellaro also had difficulties as a cadet at West Point but still admires and respects the USMA. "I personally had issues with hazing that resulted in an honor investigation and cadets being dismissed from the academy. While I was disappointed with upperclassmen at the time, I resolved myself to survive the honor investigation. I spent a year in the honor mentorship program and became an honor representative as a Cow and Firstie to protect other cadets from false cases and hazing in the future."[63]

Although Tony saw lapses in judgment and behavior by cadets in his leadership cadre, he believes that West Point's leadership handled all serious lapses in behavior with appropriate investigations and judicial punishments. He believes that his military training was excellent. The vast majority of TAC officers and NCOs gave generously and selflessly of their time and energy.[64]

Ross Pixler reflected that "while at the Academy I saw everything I did not like about the Academy. I saw all of the flaws up close and personal. I felt personally slighted by many actions and behaviors and at the time would attribute them to systemic or acute problems based on my perception at the time."[65] But today Ross's perspective has changed, and he does not see these difficulties in the same way. "My perception of fair is someone else's perception of iniquity. You will never eliminate all problems in an institution."[66] Although his experience at West Point was imperfect, he does not believe that during his time there or today any other service academy, civilian college, or university "offers the same level of opportunity, mentorship and leadership development."[67]

Looking back, Bobby Sickler did not see any serious lapses at West Point while he was a cadet. He never saw anything approaching systematic dereliction of duty, as some critics of the USMA have alleged. As he recalled, "The foundation West Point laid in the early years of my adulthood has been a significant contribution to any success I have enjoyed in life since."[68]

Conclusion

The central argument of this chapter has been that West Point is always striving to be a citadel of America's most honorable civic and military values. Like any human institution, it is imperfect because it is run by and attended by people who are themselves flawed. When the USMA has stumbled and occasionally fallen as every human institution does, it has not hidden behind damage control and public relations spin. Instead, the USMA itself has made the problem public and then confronted it with unflinching candor, courage, and determination. In short, West Point is always in the process of becoming something better, something closer to its ideals.

To grasp how long and hard that process has been, it is important to widen our lens and return briefly to the 1970s and the 1990s. West Point's response to two major international events helped drive agonizing debates within the USMA that in turn produced some of the important reforms we have just reviewed: the American defeat in Vietnam and the end of the Cold War.

The U.S. defeat in the Vietnam War was especially traumatic for the Army and for West Point. For the first time in American history, the U.S. Army had lost a war. Within the Army and within West Point there was turmoil and upheaval. There were so many questions to be addressed. Why had the U.S. Army been defeated? What could explain the Army's ethical lapses, corruption, and moral bankruptcy? As the Army's USMA was tasked with producing leaders of character, this was especially troubling because so many of the Army's leaders throughout the Vietnam war were West Point graduates. This hit the USMA with special force when its superintendent, Major General Sam Koster (class of 1942), was forced to resign in 1970 because he and thirteen other officers were charged with trying to cover up the horrific My Lai Massacre than had taken place in March 1968. Although criminal charges against him were eventually dropped, the Army's official investigation found that General Koster had failed to properly investigate the reports of the massacre. At West Point, General Koster became a symbol of the Army's moral bankruptcy in Vietnam.

General Koster's resignation left the USMA stunned and demoralized. In the years that followed, over thirty members of the military faculty resigned their commissions. There were many painful meetings at all levels of the USMA to try to come to terms with what had happened in Vietnam and how West Point should rebuild the officer corps. This agonizing reappraisal helped fuel the post-1977 reforms that in turn reshaped the moral, military, and intellectual caliber of the cadets, TAC officers, and junior military faculty who rose to become the Army's leaders. Slowly and painfully, the USMA rebuilt itself.

The second major external event that drove the reforms was the end of the Cold War, which raised serious questions about what the future role and purpose of the Army would be. These questions prompted an intense debate within the Army and the USMA between two contrasting visions of the future of warfare that took place throughout the 1990s. The first vision was that the future of warfare would look much like the past: nation-state versus nation-state, force on force, prolonged heavy, conventional warfare. The second vision of future conflict was that it would be very different from the past, encompassing peacekeeping, humanitarian interventions, counterterrorism, and, as was the case in Operation Desert Storm, speedy conventional operations against technologically inferior forces.

Within the Army and West Point, the conflict was heated. As Col. Jay Parker (Ret.), then a senior faculty member, recalled,

> Some vocal cadets, faculty and senior Army leaders often parroted the mantra that "we fight and win the nation's wars" as a way of saying we don't do all the other missions. That mattered because pockets of resistance in the Army, in Congress, and in the alum ranks were adamant in their resistance. Add to that the small but very determined band of influential individuals who wanted to reverse the changes in cadet development, halt the IAID program, turn back curriculum changes, and even reverse the admission of women, the increase of minorities and the mid-1960s doubling of the size of the Corps.[69]

It was a constant and at times fierce struggle between reformers and their opponents. In the end, the reformers prevailed and to good effect. So, when the five elite souls arrived at West Point, they entered a USMA that had undergone extensive reform and was well placed to prepare them for the wars in Iraq and Afghanistan in which they would all serve with distinction. But there were still problems to be addressed. In a rapidly changing America and a rapidly changing world, West Point still faced a formidable task: the dual mandate of building leaders of character and then teaching them to understand and successfully meet the challenges of that profession in the complex counterinsurgency wars of the early twenty-first century. Going forward, as the world continues to change in unforeseen ways, West Point will always find it necessary to adapt its programs and policies in order to prepare its graduates for the world of tomorrow, the world in which those graduates will live and lead, not the world of yesterday.

7

NICK ESLINGER

R-Day

On June 30, 2003, Nick Eslinger, a graduate of the USMA's preparatory school, arrived at West Point. As a prepster he had a big advantage coming into the notoriously harsh Beast Barracks and the tough Plebe year. Beast Barracks was and is six weeks of mental and physical purgatory designed to test the mental and physical resilience of the new Plebes. Unlike the other new cadets coming straight from high school, Nick did not feel the typical R-Day anxiety. He knew the fundamentals of military custom and courtesy. The USMAPS had already trained him in the basic principles of soldiering: field craft, marksmanship, and land navigation, among other skills. West Point's prep school had also taught him how to be a cadet and what it is to be in the Army. He had exceptional military mentors invested in his success who not only inspired and motivated him but also helped him gain a deeper understanding of the Army from a soldier's perspective. In addition, the USMA constantly sent officers down to the USMAPS to help candidate cadets understand what was going on at West Point. As a result, Nick had already begun to develop into a soldier and a leader.

This enabled Nick to focus on helping his squad mates, and he quickly became an informal leader. If they needed to learn how to tie their gear down or needed any kind of basic military guidance, members of Nick's cadet squad quickly learned to ask him.

Beast Barracks I was all about physical training on West Point's hilly campus. Nick was well prepared. He was an unusually gifted athlete who was named Freedom High School's athlete of the year in 2002 and captained its football and golf teams. He had also had the benefit of a year of rigorous physical training at the USMAPS.

In Beast Barracks II, Nick was lucky to have an outstanding platoon leader, Cadet Lt. Nate Smith. To Nick and everyone in his platoon, Cadet Lieutenant Smith was exactly what they thought a young West Point cadet leader should be: tall, charismatic, and fit with impeccable bearing. Cadet Lieutenant Smith not only told Nick and his platoon mates that he cared about them and wanted them to succeed but also led from the front, guiding the young Plebes through every military activity required of them in Beast Barracks II. Nick remembered Cadet Lieutenant Smith with great affection and respect. "He was a quiet professional, who let his presence speak for itself. I admired him and I tried to model myself after him for the next four years."

Militarily, Nick was off to a great start at West Point. Academically, it was a different story. In his first semester, he struggled. He did well in history and English and excelled in military science and military-related PE courses such as gymnastics but barely got by in general chemistry, mathematics, and computer science. In the spring, Nick excelled in military science and military-related PE classes (including survival swimming and boxing) and did well in English but struggled in chemistry, psychology, and even history, which he had always loved.

Plebe year is notoriously difficult. Years later Nick reflected on his experience:

> The first-year experience is really psychological. The system has a way of diminishing your human desire for influence; you don't have any influence as a Plebe. You are literally to be seen and not heard, and

you only speak when spoken to. And when you see things that are going wrong, you can't really speak about it because it's not your place. That's hard going from high school, where you had a lot of influence as a team captain. I'm speaking for the general cadet here who is well known in their community; a big fish in a small pond, you are now a small fish in a big pond as a Plebe. Because of the length—the full year of feeling that—it really sobers people's egos, and you start your next year with a sense of appreciation where you can empathize with the Plebes. This makes you a better team leader because that is the next step. Your leadership development at West Point in that first year is to become a team leader of one or two Plebes in the second year.

This system has a purpose and a positive effect: it gets cadet egos under control and helps them develop a deeper understanding of the West Point experience. It also helps them develop greater empathy with the Plebes they will help lead in their second year. In turn, this empathy has a greater importance: when cadets graduate and become platoon leaders, they are better able to inspire and motivate their soldiers.

As Nick stated, "When you are commissioned as a platoon leader and you see your privates, you empathize with them because they are experiencing something very similar to what you experienced as a Plebe. And where else are you going to get that experience other than the year you spend at West Point being essentially a private? It's very important to an officer's development, I think."

Since he became a cadet, Nick knew that he wanted to make the Army his career. So, he wanted to invest in his career by majoring in military science, a subject that would contribute to his success as a platoon leader by teaching him technical and tactical competence, thereby building a foundation for every other position that followed. Nick was also inspired to be an infantry officer by the military faculty in the USMA's Department of Military Instruction. They were not only officers of character but also exuded dynamic leadership and were deeply invested in developing cadets to be effective platoon leaders.

In the summer of 2004 Nick returned to Camp Buckner, this time for Beast Barracks. He recalled that he had "a great time at Camp Buckner and really enjoyed learning about the different branches of the Army." He also stated that "I vividly remember the 'long-range patrol,' and it was my favorite training that summer. It was a tough lane, and many of my platoon members were struggling to keep moving under weight and duress, but I loved it."

For Nick, the summer of 2004 was significant for two reasons. The first was his discovery that he really enjoyed being pushed to the limits of his physical and mental endurance and wanted more. The second reason the summer was significant for him was that he decided that he wanted to join the 101st Airborne Division. He made this decision because he was blown away not only by the phenomenal quality of the officers and NCOs in the battalion from the 101st assigned to train his class but also the way the officers led through their NCOs.

That summer Nick also spent five days at Fort Bragg, North Carolina, in the Special Forces Individual Terrorism Awareness Course. West Point chose him for this course after a successful tryout in the spring. The course was designed for U.S. military personnel, diplomats, and officials from other government agencies. Its goal was to teach them how to make themselves hard targets for a terrorist in a hostile foreign country. The course trained Nick and the other personnel how to recognize if they were being surveilled, how to inspect their cars for bombs, what to look for, and how to react in foreign countries. As part of the program, Nick even took an offensive driving class in which he learned to escape hostile pursuers and ram their vehicles in case he was ever surrounded by them.

In the fall of 2004, Nick was back in the classroom at West Point. There, once again, he excelled in his military science, military development, and military-related physical education classes. He did well in German but struggled in the physical sciences, mathematics, computer science, and psychology. In the spring of 2004, the pattern was similar. Once again he excelled in all his military classes and improved in mathematics, but he struggled to maintain average grades in physics, philosophy, and economics.

The summer of 2005 brought Nick his first group leadership opportunity. He was selected to be the first sergeant for 2nd Company during the first three weeks of Beast Barracks at Camp Buckner. Nick excelled, winning the Best First Sergeant award. With characteristic modesty, he attributed his success to a "fantastic" TAC NCO who coached, taught, and mentored him in his duties as first sergeant. Nick also benefited from excellent leadership in the Cadet Cadre at the company commanding officer (CO) and executive officer (XO) levels. He learned much about the role of the NCO in the Army especially in planning, preparing, and executing company-level training.

Returning to the classroom in the fall of 2005, Nick found his sea legs academically, making the dean's list for the first time, with excellent grades in the history of unconventional warfare, English, and the foundations of engineering in addition to his usual outstanding performance in military science, military development, and military-related physical education. Unfortunately, the spring semester of 2005 saw Nick's academic performance slip, and he failed to make the dean's list because he struggled in an information technology class and, unusually, in a military science class on tactics.

For Nick, the summer of 2006 brought promotion in the cadet leadership ranks as well as Airborne School and an exciting CTLT with the 173rd Airborne Brigade in Vincenza, Italy. He began the summer serving as the summer garrison regiment command sergeant major. In this role, he was responsible for the accountability and discipline of all the cadets who stayed at West Point for summer school remedial classes. This experience taught him a lot about peer leadership as well as how to influence and inspire cadets with low motivation. Next came Airborne School at Fort Benning, Georgia, where Nick had a blast. He loved the training, the jumps, and the interactions with the junior enlisted soldiers who were interested in cadet life. The high level of professionalism of the airborne cadre left an indelible imprint on Nick, reinforcing his ambition to join the 101st Airborne Division.

Nick's CTLT in Italy was in his words "an incredible experience." The 173rd Airborne Brigade was a coveted slot for cadet CTLT, and

Nick felt very lucky to be selected. The USMA's choice reflected his outstanding military leadership record. He flew to Italy the same day he graduated from Airborne School and shadowed a platoon leader in Chosen Company, 2nd Battalion, of the 173rd Airborne Brigade. Once again, Nick was lucky. The 2nd Battalion's commander was Lt. Col. William (Bill) Ostlund, an inspirational leader widely regarded as one of the finest officers in the U.S. Army and known for his commitment to training and preparing his soldiers for combat, his personal courage, integrity, and passionate commitment to his troops. Within a couple of years, Lieutenant Colonel Ostlund's battalion would become the most decorated unit in the Afghanistan War. Nick joined Chosen Company during a major training event in which the battalion traveled to the NATO air base in Aviano, Italy, to conduct airborne operations and platoon-level training. As Nick recalled, "It was truly eye-opening to see the difference between jumping out of airplanes with other airborne students versus jumping out of airplanes in an actual tactical battalion. I got to jump four times and was put in charge of patrol base operations during our last mission. The NCOs really encouraged me and helped me where I was weak. I learned a ton from the platoon leader I shadowed, and after CTLT I knew from that point on that I wanted to be an infantryman, specifically an airborne ranger."

There is one poignant footnote to Nick's CTLT experience with Chosen Company. A young first lieutenant, Matthew Ferrara (West Point class of 2005), became a mentor to Nick and spent a lot of time with him asking questions, learning about Nick's life and his cadet experience as well as his career goals. First Lieutenant Ferrara also gave Nick a lot of useful advice. They worked out in the gym together almost every morning. In Afghanistan the following August, First Lieutenant Ferrara received the Silver Star for his heroic defense of his combat outpost in Aranas, Afghanistan, when it was partially overrun by over one hundred Taliban insurgents and half of his platoon was wounded. To defeat the Taliban attackers, First Lieutenant Ferrara called in A-10 Warthog close air support aircraft to strafe within ten meters of his own position. Later in November 2007, he was tragically killed when his platoon was ambushed in the mountains of Afghanistan as he tried to reestablish relations with the elders of Aranas.

Fresh from a highly successful summer command at West Point and CTLT with the 173rd Airborne Brigade with an A+ grade, Nick returned to the classroom in August 2006. That fall his academic performance soared, and he made the dean's list. In military science he took a course called "Combat Leadership" that left a lasting impact. Taught by Col. Pat Sweeney, an experienced combat veteran, the course was an intimate analysis of combat. Part of the course focused on learning about how to deal with combat stress and fear. Here, Colonel Sweeney taught Nick that combat stress is always there regardless of whether you are being shot at or blown up. The threat itself can weigh on a soldier. Through assigned readings, small-group discussions, and Colonel Sweeney's experience dealing with his own fear and that of his soldiers, Nick learned how to be aware of and help manage his platoon's fear. In particular, he also learned the value of combat stress teams and combat counselors in diagnosing and treating a struggling soldier's anxieties and fears.

In the spring of 2007 Nick continued to excel. Taking six courses in his military science major (including operational warfighting, platoon leader responsibilities, counterinsurgency operations, comparative military systems, and colloquium in military affairs) he earned nothing but A's and B's. On May 26, 2007, Nick graduated on the dean's list with a bachelor's of science degree in military science.

Graduation Day

For Nick, graduation day, May 26, 2007, was deeply emotional. He recalled that "I cried on my graduation day, and I'm not a crier; I was blindsided by the emotion. It was joy, and then I went into the stands and I hugged my mom. It was like we made it together and we did this and she helped me get across the finish line. Two seconds later, my cheeks were wet and I didn't know where it came from. I was blind-sided because everything just came to a point emotionally. It was just such an accomplishment and it manifested itself through tears, and they were joyful tears."

For Donna, Nick's mother, this was and would always be a trea-sured moment. Immediately afterward, Nick and his extended family repaired to a spot on a grassy hill near Michie Stadium where the com-mencement ceremony had taken place. Donna had a special gift for Nick. During the four years he had been away at West Point, she had lovingly compiled an album of childhood photographs that she now presented to him. Slowly they went through each page together, paus-ing for a long, loving look at Donna's favorite picture: a photograph of Nick at age four in camouflage pajamas. Beneath the photograph Donna had written "My Hero."

Years later, Nick reflected on the value of the USMA academic pro-gram. West Point's rigorous academic program taught him valuable skills that helped him be a better officer. Nick recalled,

> I struggled academically, it was a definite challenge for me. And through that experience I learned a lot about my tendencies. I am a procrastinator; I am a crammer where I will cram for the test versus try and just prepare over the course of time. But it is at West Point that I learned that about myself and adapted to be less of a procras-tinator and less of a crammer so that as an officer, when outcomes matter—not just to me as a student but to me as an officer that affects my platoon—I wouldn't be up against a deadline. I could get out in front of things; I could produce a better-quality product, and I wouldn't necessarily have learned that if I didn't have the demands that West Point academics put on me.

Militarily, by the time Nick graduated from West Point, he had become an exceptionally well-prepared, caring, and effective leader.

West Point had given him great leadership opportunities within the cadet cadre and in the field army. Thanks to his Plebe experience and his inspirational military mentors at the USMA and in the field army, Nick understood the needs of twenty-first-century American soldiers and how to inspire and lead them.

8

TONY FUSCELLARO

R-Day

Tony Fuscellaro arrived at West Point for R-Day on July 2, 2001. His Beast Barracks experience was "taxing but not terrible." Later, Tony remembered,

> I enjoyed the physicality of it and the workouts. I thought the actual low-level tactical training was interesting. I was certainly inspired by the challenges there, and I think I've always been the kind of person who thrives when I am fully taxed. So, when there is too much to do, that's when I really buckle down and achieve my best results. And that was something that Beast does very well. It fully taxes you, and I distinctly remember it being very difficult but always seeming to find enough time to do the things that I had to do. It brought a greater focus out and I appreciated that.

Unlike Stephen Tangen, Tony's Beast Barracks squad enjoyed a high-caliber cadet cadre whose members were interested, motivated, and sincere in their attempts to train Tony and his fellow Plebes. In his squad, there was no verbal abuse, much less hazing. This cadet cadre not only took seriously the idea that they were the latest link in the chain of the USMA's storied "Long Gray Line" but had also fully absorbed the inspirational servant-leader model that Lt. Gen. Dave Palmer, Brig. Gen. John Abizaid, and Lt. Gen. Dan Christman had conceived and implemented. They gave Tony's basic training their all: they were "demanding but never demeaning." However, outside of his own Plebe company, Tony was aware of a small minority of the cadet cadre who never absorbed the USMA's reform agenda and continued to verbally abuse and haze Plebes.

Tony was just beginning life as a cadet when the 9/11 terrorist attacks occurred. Tuesday, September 11, 2001, was a perfect early fall day in the Hudson Valley, north of New York: warm without a cloud in the sky. The terrorist attacks in New York and Washington, D.C., shattered that idyllic scene. When Tony first learned of the attack he was in his Plebe English class in Thayer Hall, a granite fortress-like building embedded in the cliff above the Hudson. The first hint that something was wrong was the commotion in the hallway outside the classroom door. After trying unsuccessfully to keep his class on task, Tony's English professor went outside to find out what was going on. After consulting other professors teaching in nearby classrooms, he came back in and turned on the television. That was where Tony and his classmates saw the second airliner hit the World Trade Center. Tony remembers Thayer Hall becoming eerily quiet: "I mean, it was starkly quiet. At that point, I remember the instructor telling us 'okay, everybody is going to stay here until we figure out what's going on.'"

Outside Tony's classroom there were a lot of discussions about what to do with the cadets. No one knew whether West Point was a target, and there was concern that if the cadets were brought together in the mess hall, they would offer a tempting target for terrorists. The superintendent, Lt. Gen. William Lennox, therefore sent the cadets back to their barracks, which were dispersed throughout the central post. The tactical officers for Tony's company came through the barracks trying to reassure the cadets. Their message was that "this is a significant event.

We don't know yet who is attacking us, or whether West Point is a target, but we are all going to stay here together until we ride it out." Recalling the day's events many years later, Tony said, "It was just a shocking day, a day I'll never forget. Our senior leaders, our mentors, told us, 'You are now part of an army at war. Everything you are doing here at West Point is now becoming more real. You are going to be tested. Your class is going to be tested.' And I remember the impact that it had on me. It certainly made me focus personally more on my military duties at West Point, more on my military education than on academics." After 9/11, Tony's priority was to do everything possible to develop into an effective platoon leader.

For Tony, his classmates, and everyone at West Point, the terrorist attacks of 9/11 had given everything they did a greater sense of importance and urgency. As an instrument of national policy, the U.S. Army had overnight become rapidly relevant. Immediately, Special Operations Forces were in action in Afghanistan, and preparations for a wider ground war were under way. Already, Tony could see some young military faculty leaving the USMA to deploy. Throughout West Point, there was a building sense of urgency every day, every week, every month. Tony and his fellow cadets felt they were coming closer to becoming platoon leaders in combat. Within two years, Tony would witness the arrival of combat veterans who would teach him military science infused by the hard-earned lessons of war.

In his Plebe year, Tony had another defining moment at West Point: a violation of the honor code that almost cost him his place at the USMA. The incident was a result of innocent youthful exuberance. Tony and his fellow Plebes in Company F1 carried out what they called "spirit missions," capturing Yearlings on their birthdays especially those who had been stern taskmasters during daily Plebe duties. Tony's company was especially keen on spirit missions and had captured almost every Yearling who supervised them. The Yearlings got fed up, and there was a fight in which Tony got a black eye.

The next day Tony's company sergeant asked what had happened. Tony lied, saying he had walked into a closet door. She didn't believe him. "I'll talk to you later," she warned. Because Tony was a Plebe in his first semester, his company sergeant was trying to give him another opportunity to tell the truth. Tony was a young man of integrity but was conflicted, torn between the norms of the blue-collar culture in which he had grown up in his close-knit community and those of West Point. In this situation, Tony was torn between the values of a culture whose cardinal tenet was "always protect your family and friends" and the strict requirements of the cadet honor code: "No cadet may lie, cheat, steal, or tolerate those that do." As Tony put it, "Snitching was not something I was trained to do." Later that day in the mess hall, his company sergeant pressed him again. She threatened that if he didn't tell the truth, she would refer the matter to the USMA's honor committee. Once again Tony lied, and the company sergeant turned him in because it was her duty to do so.

Now Tony faced a full honor committee investigation that could lead to his expulsion from the USMA. The committee was lenient with him because he was only in his first couple of months as a cadet and had not yet had the time to fully absorb and internalize the USMA's honor code and because he was only trying to protect the other cadets in a fairly innocent escapade. The committee also saw how he had been struggling to reconcile his two conflicting concepts of integrity. Instead of expelling him, the committee gave Tony a second chance by assigning him to the cadet honor mentorship program. This is a one-year program in which a cadet has to study assigned readings about what honor truly means and then discuss them in a weekly meeting with a military mentor. Tony's mentor was outstanding, and the weekly meetings proved transformative. Tony now embraced West Point's honor code with all the zeal of a convert. His transformation was so complete that he became a member of the honor committee and sat on many honor boards in his Cow and Firstie years. Later Tony summed up his experience. "I ended up really understanding and growing in appreciation for the way West Point uses the cadets to uphold their own. That was the most important developmental thing outside of classes that happened to me at West Point."

Despite the impact of the 9/11 terrorist attacks, the stress of his honor code violation, and the added time necessary to study and prepare for the cadet honor mentorship program, Tony had a successful fall semester. He earned a series of solid B grades across science, mathematics, and humanities classes. In both the fall and the spring, Tony's military science grades were stellar. In the spring, he also improved his overall academic performance despite struggling a bit in chemistry.

For Tony, the summer of 2002 brought Cadet Field Training. His Plebe year prepared him well as he had begun to develop from someone who was very independently minded into someone who was beginning to think more of his role within the team: the team was only as strong as its weakest link, and he was determined never to be the weakest link. That summer Tony thrived, learning infantry and small-team tactics.

At the same time because of good mentors at the USMA, Tony had begun to think about what branch of the Army he wanted to serve in. Cadet Field Training offered cadets the opportunity to try every branch of the Army, and Tony was keen to try everything. He fired an M109A6 Paladin self-propelled howitzer, got into an Abrams battle tank, and got a ride in a Black Hawk helicopter. The armor branch did not appeal to him because he didn't like the idea of being a tank commander inside a big vehicle when he wasn't the person driving it or firing its main gun.

The flight in the Black Hawk captured Tony's imagination. Unlike so many other military officers who have a family history in aviation, Tony had none. He was not one of those kids who grew up the son or daughter of airline pilots dying to fly. His love of flying was born that summer during Cadet Field Training at Camp Buckner.

Tony returned to West Point in late August 2002 to begin the academic year. He tackled his courses with determination and focus, generally earning solid grades across a wide range of disciplines.

Under the new Abizaid leadership development model, the fall of 2002, the beginning of Tony's second year, also brought a new responsibility: mentoring Plebes. Tony embraced the Abizaid servant-leader model with enthusiasm. He had a dynamic military instructor who taught him that it was more important to be a part of your unit rather than its leader. The instructor's message was clear and emphatic: "You're not just the person at the top of the triangle, you're the person in the center of the circle. You need to put others first. You need to lead from the front. The leader is the one in front that pulls everybody in the unit along, that shows them the direction, that motivates them and inspires them to come with him." This insightful guidance shaped Tony's approach to leadership at West Point and throughout his Army career.

The summer of 2003, Tony's third summer as a cadet, brought him two contrasting experiences: early CTLT at Fort Leonard Wood followed by flight training at the University of North Dakota. Cadets normally do their CTLT training as rising Firsties, but Tony's TAC officer had mapped this program out so he could take on a leadership role during Cadet Basic Training at West Point the following summer.

Because he was undertaking CTLT as a Cow, a junior classman, Tony was forced to take an assignment he was not very interested in at Fort Leonard Wood, Missouri, the U.S. Army's largest training base. His four-week assignment was to join an army basic training company and shadow a lieutenant and learn from his experience. It wasn't a great learning experience. To begin with, there was very little to be learned from the activity Tony was observing. Army Basic Training is very similar to Cadet Basic Training at West Point. It's about running every morning, climbing obstacles, learning about weapons, and the simple daily or weekly activity planning associated with it. Worse, there were hardly any lieutenants to shadow. Instead, Tony shadowed senior NCOs who were not particularly interested in teaching a West Point cadet.

Always an acute observer and keen student, however, Tony compared the drill sergeants who really cared about their recruits and those who didn't. The ones who cared got better results by going out of their way to encourage and help their young raw recruits master tasks such as putting together a weapon and tackling a tough obstacle course. Those who didn't really care just put a recruit on a task and, if the recruit struggled or failed, just moved on. The leadership lesson Tony drew was that many young soldiers execute tasks in an inexperienced and inefficient way and just need a little more guidance and time to master a task. Positive leadership gets positive results, just as he had been taught at West Point.

From Fort Leonard Wood, Tony went on to spend the rest of the summer of 2003 at the University of North Dakota's flight school at Grand Forks International Airport. There, he spent a month learning to fly in a small Schweizer 300, a light utility helicopter with a three-bladed rotor. Built by Sikorsky, the Schweizer is mostly used for flight training and agriculture. It is smaller and more fragile than a television news helicopter and has a tiny cockpit where instructor and student share a common control stick facing eight dials on a control panel the size of a tablet computer.

After a month, Tony's pilot instructor cleared him to fly solo. As Tony recalled, "I remember taking off and flying, being up there relying only on myself. This was something that I understood and felt right to me. That's when I made the decision to go aviation out of West Point."

After he returned to West Point in the early fall of 2003, Tony began to take courses in his major—art, literature, and philosophy—and made the dean's list in both the fall and the spring.

In his final summer as a cadet, Tony became a regimental commander at Cadet Basic Training, one of the Firsties in charge of a group of basic training companies. He thrived in the monthlong training program for the cadet cadre and enjoyed applying his vision of positive inspirational leadership to the development of the new class of Plebes over the next six weeks. Tony enjoyed the opportunity to lead and interact with the new incoming Plebe class, glad to see their confidence grow and their comfort with the USMA's rigorous program improve. As he told me later, "I took a ton out of it." His plebes did too. Throughout the 2003–2004 academic year, many of them sought Tony out for advice. His summer leadership experience at Cadet Basic Training also prepared him well for his new role as a company commander within West Point's garrison.

In his final academic year, Tony continued to earn stellar grades in military science and achieved generally solid academic grades in his academic major. He made the dean's list in both semesters.

In the fall semester of his Firstie year, like all cadets, Tony had to choose the branch of the Army he wanted to serve in. Despite his love of flying, he was torn between aviation and infantry. Like many cadets, Tony had always felt that the infantry was the heart of the army. It was not an easy choice. The way he approached it was to embrace the air cavalry spirit, the infantryman in the air, always ready to help comrades on the battlefield. So, he chose Army aviation but vowed he would make sure to end up in an aviation unit that was as close to the infantry as possible. Even as a cadet, he had fallen in love with the scout mission. Tony graduated in May 2005 with a BSc in art, literature, and philosophy.

9

ROSS PIXLER

R-Day

On July 2, 2001, Ross Pixler arrived at West Point two hours late for R-Day. His parents, Reid and Larissa, had driven him the nearly three thousand miles from Phoenix but on the final leg of the journey had misread their maps and the road signs and got lost. This was before the time of GPS. Ross arrived in a lather in Eisenhower Hall five minutes after the last group of new cadets had been admitted for processing. When he tried to join them, a cadet blocked his way. Later, Ross was admitted and went through the rough cadet induction alone. As Ross wryly observed, "I had the undivided attention of many a cadet all devoted to my professional well-being."

If Ross was late, he was also overconfident. He had spent four years in JROTC and had commanded his JROTC battalion. He knew a lot about the military and was physically very fit. Very quickly, however, Ross found that he was in over his head. Beast Barracks turned out to be "a good, fun, humbling experience," as he put it. Physically he never really felt challenged. As a mountaineer with exceptional endurance who had recently rescued his best friend Jared Sibbett in a snowstorm on Mount Orizaba in Mexico, this was not surprising. But Ross struggled with West Point's cadet culture and had difficulty memorizing "the knowledge."

During Beast Barracks I, Ross's squad leader was Cadet Sgt. Howard Lim, who was aggressive, caustic, and demanding. He made Ross's life miserable. Although Ross did not like or respect Cadet Sergeant Lim, he grew, became stronger, and developed more under him than other members of the cadet cadre. Later Ross opined, "Comfort is antithetical to change, and when you need change for growth and development, a little hardship goes a long way." Miserable though those first three weeks were, Ross believes that he greatly benefited from Cadet Sergeant Lim's harsh, demanding leadership.

In Beast Barracks II, when the emphasis shifts to basic infantry skills, Ross was in his element: there was mountain climbing, rappelling down a fifty-foot cliff, firing a rifle, and learning how to capture a hill or a trench from the "enemy" as well as night patrols under "fire" from soldiers in the field army. Ross's squad leader was Cadet Sgt. Alexander Lane. He was sincere, motivated, and demanding but never caustic or aggressive. As a former JROTC commander, Ross had a blast in Beast Barracks II.

Throughout the six weeks of Beast Barracks, Ross's biggest struggle was psychological: he never felt more alone than during Beast Barracks, and at night in his bed he was homesick. But Ross had remarkable resilience and persevered, figuring out Beast Barracks one day at a time.

Once the academic year began, Ross decided to major in nuclear engineering, a brand-new academic major that gave cadets the most broad-based engineering education. For Ross, this was a natural choice. He had had four years of science classes in high school, including three college-level classes, and had done well in them. But West Point's Ivy League academic standards presented a challenge of an entirely different order of magnitude. Ross did not excel academically and constantly struggled to be average. Nuclear engineering was one of the most difficult, demanding majors at West Point. Ross chose it not only because he liked this subject but also because it challenged him the most. And he was always up for a challenge that tested him to his limits.

For Ross, Tuesday, September 11, 2001, changed everything. It began normally: early-morning parade, breakfast, and freshman chemistry, his first class, which was in Bartlett Hall, the USMA's science building. It was a ninety-minute class, but the instructor normally let the class out partway into the second hour. That morning was no exception. Ross got out of class and went back to his room to relax before his next class. He was surprised to find his roommate, Ben Hirsch, watching television news through his computer. Ross asked, "How can this happen? How can an airline pilot crash a big commercial jet into the World Trade Center?" Ross didn't believe it was a freak accident. Instead, he suspected foul play. As they watched together, a second airliner hit the World Trade Center. Ben asked, "What are the odds of this happening?" Ross replied, "No. This is an attack. We are getting attacked by someone." Ross's fears were fully realized. After the second jet flew into the World Trade Center, Ross and Ben knew this was war. From that moment onward, everything Ross and Ben did at West Point assumed a new meaning.

Like other cadets, Ross wanted to go down to New York City to help recover the dead and injured from the rubble. USMA superintendent Lt. Gen. William Lennox said no. Cadets should carry on with their education and training so as be fully ready to fight the war that had just begun. For Ross and the other cadets, the next three years were an anxious and frustrating period. They felt trapped in West Point classes when they wanted to be in combat; they worried that the war would be over before they could join the fight. When Ross learned that during World War II the USMA had graduated whole classes a year early, he and his friends hoped the same would happen to them. But it didn't. What did happen was a new sense of urgency coupled with a new realism in their education and training. In Ross's military science classes, everything refocused on Afghanistan, and later Iraq, rather than hypothetical conflicts in imaginary countries.

Academically, the 2001–2002 school year was a difficult adjustment for Ross. He did well in military-related physical education, mathematics, and chemistry but struggled in the humanities and the social sciences. This pattern continued in the spring semester of 2002, but Ross had difficulties with more advanced courses in chemistry and mathematics.

In Ross's mind, 9/11 had changed everything. He knew that he was now preparing to do something bigger and more important than peacetime service. For Ross, there was no greater way to contribute than to serve the country in a time of war than by leading troops in combat. Even before he went to Cadet Field Training in the summer of 2002, Ross knew that he would forever regret selecting any branch of the Army but infantry.

In the summer of 2002, Cadet Field Training offered Ross the opportunity to try out every branch of the Army. He was predisposed toward infantry but was also interested in armor. However, when he saw the difference between the leadership responsibilities in armor versus infantry, his choice was easy. Armor meant leading a platoon of tanks with a relatively small number of soldiers who engaged the enemy from a distance; infantry meant leading a much larger number of soldiers and engaging the enemy at close quarters. Ross chose infantry.

The academic year 2002–2003 was once again challenging for Ross. The fall semester saw his most consistent overall academic performance, but the spring semester was a very difficult one. He excelled in military science but struggled to maintain average grades in his academic disciplines, especially in engineering and mathematics.

What helped sustain Ross during these challenging years was the friendship and mentorship of his "second set of parents," Dr. Brian Moretti and wife, Betty. Ross knew Betty because she was his dental hygienist. On Thanksgiving morning in 2002, Ross got on the Metro North commuter train at Garrison Station, just across the Hudson from West Point. He was going into New York and then on to New Jersey to spend Thanksgiving with his new girlfriend's family. As he was settling into his seat, Ross took a telephone call from her telling him not to come because she had decided to end their relationship and get back together with her former boyfriend. Ross was devastated. He returned to West Point because he had nowhere else to go. But when he got back to West Point, the USMA had shut down for the holiday, and the barracks were closed.

Luckily, Ross ran into Betty Moretti outside the USMA's dental center. She told him that he was spending Thanksgiving with her, her husband, and their two sons and other cadets at their redbrick Victorian home on "colonels' row" overlooking the Hudson. Their friendship flourished, and they became Ross's sponsor family. Later Brian and Betty became godparents to his children. Colonel Moretti and his wife liked Ross. They found him to be "just a nice kid; quiet, reserved but genuine and caring." They could see that "the goodness of his character was already established."

A deeply religious family, the Morettis could also see that Ross was someone for whom faith and family were of paramount importance, someone with a clear and strong moral compass who always did his best to live the values he professed. Ross was an active member of the Catholic congregation at West Point and almost never missed mass. As a teenager who had felt an almost religious calling to military service, it was not surprising that he once described himself to the Morettis as "a soldier of Christ." By that he never meant that he was some kind of twenty-first-century version of a medieval Crusader; rather, Ross meant that he was a soldier suffused with a strong Christian ethic, a deep commitment to doing what was morally right.

Over time, Ross's personal warmth and generosity helped deepen his friendship with Brian and Betty, and he became a member of their family. Every week when he came to visit the Morettis, his first question was always "what do you need?" His last question was always "can I help you clean up?" Betty recalled that "I treated him like I would one of my sons." The relationship deepened further when Ross began taking nuclear engineering classes taught by Colonel Moretti.

The 2002–2003 academic year brought more challenges. The fall semester saw Ross maintain his cumulative GPA with a solid overall academic performance, but the spring semester was especially difficult. Ross excelled in military science but was always an adventure seeker, and as a Yearling in the 2002–2003 academic year, he passed the two-week-long prescuba course and was accepted into the Army's elite prescuba combat diver school in Key West, Florida. Unlike Stephen Tangen, Ross had never done much swimming. So, giving up his spring break to test himself to his physical and psychological limits was an extraordinary thing for a nineteen-year-old to do. It was also eloquent testimony to his mental and physical resilience. In Ross's drive to succeed in prescuba combat diver school, Colonel Moretti compared him to Sir Edmund Hillary, the conqueror of Mount Everest who once said, "It is not the mountain we conquer but ourselves." For Ross too, it was about self-mastery.

Prescuba was brutal. The rule was that when underwater you could not come up for air without the approval of the instructor. If you did, it counted as quitting and the instructor would drop you from the course. The rules did, however, allow students to have a couple of shallow-water blackouts without washing out of the program. After one blackout, Ross had to be admitted to the hospital because he was vomiting blood. Within a few hours, he was back in the pool. Later Ross recalled that combat diving school was "so difficult" and "so torturous" that it caused him nightmares for years afterward. Despite the extreme difficulty, he passed this brutal, physically demanding course but never had the opportunity to complete the more academic and hands-on portion of the combat diver qualification course. The reason was that many of the instructors had been redeployed to Afghanistan or Iraq, and only eight candidates were allowed to complete their training at Key West.

Ross spent the summer of 2003 as a cadet first sergeant for Cadet Field Training at Camp Buckner, where he excelled as a leader. Cadets have a range of choices of Army specialty schools for summer training, and Ross had not yet made a choice. With typical determination and grit, he chose the Army Assault School, one of the toughest. As its website describes the experience, "U.S. Army Assault School is a 10-day course designed to prepare Soldiers for insertion, evacuation, and pathfinder missions that call for the use of multipurpose transportation and assault helicopters."[1]

Because trainees are often doing dangerous tasks under intense pressure, "a successful candidate must possess a keen eye for detail and a

dedication to meticulous preparation."[2] Ross had no difficulty with the
first combat assault phase of the course. In it, he learned the basic principles of aircraft safety, air-assault, pathfinder, and medevac operations. He
also passed phase II, "Slingload Operations," tethering heavy loads of up
to eight thousand pounds to the bottom of a helicopter. Inspecting the
helicopter's load requires the trainee to concentrate intensely and be as
precise as possible. That was Ross. The final phase of Air Assault School
was tailor-made for a mountaineer such as Ross. Here, students must
be able to successfully complete "two rappels from a 34-foot tower and
two from a UH-60 Black Hawk, hovering between 70–90 feet."[3] Before
graduation, students "must also complete a 12-mile foot march in full
gear plus a rucksack in less three hours." Ross was in his element and
accomplished the rappelling and the foot march with ease.

In 2004 between his Cow and Firstie years, Ross had to do his CTLT.
Beforehand, he left West Point to take a three-week AIAD internship
with NASA at the Early Flight Fission Test Facility, a part of the Marshal
Space Flight Center in Huntsville, Alabama.

Funded by the USMA's Association of Graduates, the AIAD program is designed to enable cadets to apply concepts learned in their
coursework to real-world settings. For those such as Ross who have
already chosen majors, it is a great opportunity to develop their knowledge and skills in their chosen field. For those who have not yet chosen
majors, AIAD aims to show them what advanced work in an academic
discipline looks like.

Working with NASA that summer was an incredible opportunity for Ross, and he learned a lot. He joined a team working on the
development of the next-generation space shuttles testing the viability
of powering them with nuclear energy. Since his degree was in nuclear
engineering, he found this a deeply rewarding experience. Ross so
impressed the NASA team that he was offered a job once his five-year
Army service commitment was up, but he declined, choosing instead
to stay in the Army.

Ross spent the remainder of the summer of 2004 on his CTLT
with the 7th Special Forces Group (Airborne) at Eglin Air Force Base,

Florida. He was assigned to a twelve-man operational detachment, the cornerstone of every Special Forces company and battalion. At the time, the operational detachment he was attached to was going through the Special Forces Advanced Urban Warfare Course, so he deployed with them to Nevada, New Mexico, and Texas for training, force-on-force, and live-fire exercises. Ross was unimpressed and uninspired by some individual soldiers in this operational detachment because in his view, they lacked the moral character he had expected. It was not the learning experience he had hoped for.

In late August 2004, Ross returned to West Point for his final academic year. It was to be his best. His fall semester saw a strong overall performance with a solid B average marred only by a near failure in a mechanical engineering course. In the spring semester, Ross found his academic sea legs, earning him a place on the dean's list thanks to strong performances in his nuclear engineering, military history, and constitutional/military law classes.

Ross graduated in May 2005 with a degree in nuclear engineering. For any cadet, the four-year journey through West Point is remarkable. In four years, cadets develop and mature while their leadership qualities begin to blossom. In Ross's case, he had grown from a quiet, reserved, anxious young Plebe into a confident leader whose goodness of character had been powerfully reinforced by West Point's honor system.

With singular determination and focus, Ross worked on developing his ability to become the best possible infantry platoon leader he could be. He greatly benefited from his studies in West Point's military science classes, including his assigned reading, the insights and inspiration of his military instructors, and especially the intense seminar discussions with these combat veterans. From them, Ross learned the characteristics of successful officers: how they inspire and lead soldiers in combat and how they set and achieve high standards.

Four members of his cadet cadre also inspired Ross: Amos Camden, Riley Bock, Dan Whitten, and Laura Walker. Tragically, all four were killed in combat in Ross's Firstie year or soon thereafter. These deaths took a personal toll on Ross because he respected these people so much and was so close to them personally. He named his son Riley Camden Pixler in honor of his friend Amos. But not even their deaths could weaken Ross's resolve to be the best platoon leader he could be.

Although Ross had struggled academically to maintain average academic grades, he left the USMA very well prepared for his further military training at Fort Benning, Georgia, and leading troops in combat. As Ross went through West Point, the most beneficial moments for him were upperclassmen speaking authoritatively, rather than screaming at him, in "a calm, quiet voice in my ear" advising and encouraging, telling him how to execute a task more efficiently. Ross's experience proved that the West Point "Changes" could produce outstanding officers well equipped to bring courageous, innovative leadership to an Army facing new challenges in Afghanistan and Iraq.

10

BOBBY SICKLER

R-Day

At West Point, R-Day July 2, 2001, was a beautiful early summer day. Bobby was nervous. He was seventeen, about to begin a major life-changing experience from a rural West Virginia high school over-achiever to a West Point cadet. Like the other elite souls, Bobby had read everything that had been published about West Point and the cadet experience there. But unlike the other elite souls, he had two advantages born of his long military heritage. The first was that his father had served as a military officer and was well placed to give him nuggets of sound advice, including "don't ever give up" and "the cadet cadre don't actually hate you. It all has a purpose." Bob Sickler had also told his son many stories about his experiences at Marine Corps Office Candidate School at Quantico during the Vietnam War, likening them to the boot camp sequence in the movie *Full Metal Jacket*. Bobby didn't expect Beast Barracks to be as bad as that, but he thought it wise to be prepared for the worst. The second advantage that Bobby had was that he had experienced a taste of Beast Barracks at a summer camp at the USNA. For a few hours, the USNA treated Bobby and the other summer campers to an enhanced version of Beast Barracks much worse than anything he was to experience at West Point. Instead, his Beast Barracks was "demanding but not demeaning."

As was the case with his classmate Tony Fuscellaro, Bobby's cadet cadre at Beast Barracks was seriously motivated to help Bobby and squad mates succeed. They had fully absorbed the inspirational servant-leader model and implemented it with enthusiasm and intelligence. The first three weeks of Cadet Basic Training take place on the USMA's main campus and consist of intensive physical training. As an experienced and natural middle-distance runner, Bobby found the running easy. The push-ups and sit-ups were more of a challenge. Thanks to years of running and physically demanding farm work, Bobby was fit but just not the type of fit West Point required. Beast Barracks I was really tough for him. Although some of his fellow Plebes considered quitting, Bobby—like the other elite souls—never did.

By contrast, Beast Barracks II was demanding but fun. Thanks to his father, Bobby had long absorbed the core values of military culture. As a result, he had always wanted to be a soldier and really enjoyed being in the company of other young men and women who also loved the Army and wanted to be officers. In addition to mountain climbing and rappelling, Beast Barracks II teaches cadet basic infantry skills and ends with a fifteen-mile march from Camp Buckner to West Point. Bobby was in his element in the mud and rain and passed Cadet Basic Training.

For Bobby, September 11, 2001, began normally: morning formation and then breakfast. His first class was a computer science lab in Thayer Hall taught by a Navy reservist. During class, an Army officer came in and whispered to her that something terrible had happened. Captain Schwartz was visibly shocked by the news. One of Bobby's classmates asked her what had happened. She replied that there had been a bad accident in New York City. For now, she would continue on with class but would be glad to discuss it before the end of the period. Later, another Army officer came in and told Captain Schwartz that a large airliner had crashed into the World Trade Center. She shared the news immediately with Bobby and his classmates.

Like many cadets, Bobby thought this was a plane crash, not a terrorist incident. He remembered that a small Cessna had crashed into a New York City skyscraper a few months earlier and thought this was another aviation disaster caused by pilot error or mechanical failure. When the cadets left class, Bobby found what he later described as "a weird atmosphere" in the corridor. As he walked down the hallway to his next class, someone had turned on the television in the Thayer Hall auditorium. Bobby had a few minutes before his next class, so he sat down and watched the shocking TV coverage. Among the cadets, there were rumors that the USMA was going to cancel classes. At lunch formation, however, the superintendent, Lt. Gen. William Lennox, told the assembled cadets in Washington Hall that classes would not be canceled because at this point the cadets' best contribution to the conflict that was now beginning was to go to class and prepare themselves to become officers to lead their troops in what would soon become known as the global "war on terror."

At first, Bobby did not internalize the idea that the United States was going to war. After all, USS *Cole* had been attacked in Aden, and the United States had not gone to war. Thus, at first he did not believe that this new conflict would affect him. As a Plebe he could not imagine that the United States would still be fighting a war in four years' time when he graduated. Col. Bob Sickler, Bobby's father, had not deployed to Vietnam but did deploy during Operation Desert Shield/Desert Storm, but for him it had been a deployment measured in months, not years. To Bobby, the idea that there would be a full-scale war still going on after he graduated in May 2004 seemed remote.

Academically and physically, Bobby found Plebe year extremely challenging. Pendleton High School in rural West Virginia was underresourced and never equipped to challenge him intellectually. As a result, Bobby had never taken calculus or advanced calculus, much less the demanding AP or honors classes that so many of his West Point classmates had. Bobby had taken precalculus as an independent study by sitting at the back of an algebra class reading a calculus book. He had a lot of catching up to do. Bobby was intellectually gifted and had coasted through Pendleton High School without doing homework or studying much. Now he had to study like never before to meet West Point's rigorous Ivy League academic standards. By the end of the fall semester, Bobby had earned a place on the dean's list with a strong GPA of 3.47, including an A– in the course "Introduction to Calculus." This was an impressive achievement for a young man with no formal preparation in calculus from an underresourced high school in rural West Virginia who was also facing other tough challenges, physical as well as emotional.

Physically, Bobby also struggled. Thanks to his work on the family farm and through his karate training, he was strong but had never had any kind of physical conditioning program in high school as so many of his West Point classmates had. As an experienced middle-distance runner, he easily met the USMA running standard but at first could not do the number of sit-ups required. Undaunted, Bobby wanted to try out for one of West Point's varsity sports teams so he could travel and get a break from the normal routines of cadet life. His father had taught him to fire a rifle accurately, and as a result Bobby was a crack shot. So, he decided to try out for the USMA's varsity rifle team. Unfortunately, the National Collegiate Athletic Association's rules required that any candidate for the team had to have shot competitively before entering the USMA. Since Bobby had not, he was ineligible. Instead, he tried out for the pistol team. His father had taught him to fire a pistol, and although Bobby was not as skilled a marksman with a pistol as with a rifle, he successfully tried out for the pistol team. As a Plebe, he found his daily practice on the firing range and his travels with the pistol team a welcome respite from the intense pressure of cadet life.

In his first semester, that pressure was made worse by a clique of upperclassmen in his cadet company who thought they should enforce standards by picking on Plebes and harassing them. This treatment was wrong and a violation of the core tenets of the reforms introduced in the 1980s and 1990s. By 2001, the overwhelming majority of cadets had absorbed and embraced the "Changes" and implemented them faithfully. But there were a small minority of upperclassmen who did not.

Bobby was appalled. Their way of treating him and the other Plebes in the company was inconsistent with the way Bobby believed subordinates should be treated. Fortunately, it was the only time in his West Point career that this kind of harassment occurred.

Bobby's strong academic performance in the fall of 2001 established a pattern for the next four years, with an excellent academic performance in mathematics and the physical sciences supplemented by strong grades in the humanities, the social sciences, and military science. Between the fall of 2001 and graduation in the spring of 2005, Bobby never failed to make the dean's list.

Summer 2002

For Bobby, the summer of 2002 brought Cadet Field Training. Led by officers, NCOs, and soldiers of the regular Army, that summer's training helped cadets learn infantry small-team tactics. It also offered an opportunity to learn about all branches of the Army. Bobby thrived learning infantry tactics, and like Tony Fuscellaro, a flight in a Black Hawk captured Bobby's imagination. It also reinforced his determination to become a pilot if he could get the new laser eye surgery to correct his vision. Bobby had learned that West Point had created a test group of the class of 2004 to have the surgery and fly helicopters. Inspired by his father, Bobby had always wanted to fly. Now he had a chance. To realize it, Bobby had to do two things. The first was to pick his summer training programs so he could get the eye surgery the summer of his Firstie year. The second thing he had to do was to earn a high enough class rank to branch aviation, and because the USMA weighted academic grades more than military science or physical education grades, Bobby gave priority to his academic classes, where he continued to excel, making the dean's list every semester.

Armed with renewed confidence, Bobby tackled his courses in the fall of 2002 and the spring of 2003 with determination and focus. Once again, he excelled in mathematics and the physical sciences, earning five

A's, including in engineering courses, and strong B's in military science, the social sciences, and the humanities.

Summer 2003

The summer of 2003 brought two fresh challenges: Army Air Assault School and CTLT. Cadets have a range of choices of Army specialty schools for summer training at the beginning of their Cow year. Army Assault School is one of the toughest. As its website describes the experience, "U.S. Army Assault School is a 10-day course designed to prepare Soldiers for insertion, evacuation and pathfinder missions that call for the use of multipurpose transportation and assault helicopters."[1]

Because trainees are often doing dangerous tasks under intense pressure, "a successful candidate must possess a keen eye for detail and a dedication to meticulous preparation."[2] Bobby had no difficulty with the first combat assault phase of the course. In it, he learned the basic principles of aircraft safety, air-assault, pathfinder, and medevac operations. Unfortunately, he failed phase II, "Slingload Operations," tethering heavy loads of up to eight thousand pounds to the bottom of a helicopter. Inspecting the helicopter's load requires the trainee to concentrate intensely and be as precise as possible. Bobby failed the hands-on inspection because he did not study as much as he should have. It was his first taste of academic failure in his life to this point. Worse, it put his West Point graduation in doubt. Bobby now had two choices: go to an easier Army specialty school or repeat Air Assault School in his final cadet summer. He chose the latter.

Bobby's CTLT assignment was not what he wanted to do. He wanted to go to an aviation or infantry unit, but the USMA gave him a four-week assignment with a strategic signals unit at Fort Dietrich, Maryland. The unit was a relay station for classified communications coming from and going to U.S. Army bases in Europe and was small, with only twenty Army personnel under a captain and some computer contractors. They welcomed Bobby, and the CO was generous with his time. But there was not much scope for the kind of CTLT experience West Point intended. However, Bobby worked with those who kept the network running and learned a lot about the Army's communications technologies. He enjoyed his assignment and learned some "neat stuff," as he later recalled. The company CO awarded him an A. Nevertheless, Bobby decided that being a strategic signals officer was just not for him: his heart was in combat aviation.

In the fall of 2003 Bobby returned to West Point, where his academic focus shifted to his major, mechanical engineering. That fall and for the rest of his academic career at West Point, he earned straight A's in his engineering classes supplemented by solid B's in all other disciplines including military science.

Summer 2004

In the summer of 2004, his Firstie summer, Bobby had a full agenda. First, he had to serve as part of the cadet cadre at Beast Barracks. Second, he had to repeat Air Assault School and finally have eye surgery. The stakes were high: if he did not pass Air Assault School the second time, he would not graduate and would not have time for his eye surgery, the key to becoming a pilot.

First came Beast Barracks I, during which Bobby was a company executive officer. He worked behind the scenes to help ensure that cadet platoon leaders and NCOs had all the resources they needed. His TAC NCO at Beast Barracks that summer, Sgt. Gill Goma, had a positive impact, teaching Bobby lessons that he carried with him throughout his Army career. A homespun "philosopher," Sergeant Goma taught him how to treat people properly, how not to worry about things he cannot control, and how to frame the world around him. "Don't worry about shit you can't do nothin' about," Sergeant Goma regularly intoned in his deep baritone.

Bobby tried his best to treat the new Plebes with dignity and kindness. On R-Day, for example, as a member of the cadet cadre, he had to sit at the head of a lunch table. His new recruits were nervous and hungry, trying and failing to be cadets. Bobby told them that "it's going to be a long day; relax, eat your food, get a full belly, you'll be ok." Nearly thirteen years later, one of those nervous Plebes stopped Bobby in the commissary at Schofield Barracks, Hawaii, where they were both serving, to thank him for his kindness that distant R-Day. The former Plebe was now a company commander in the infantry brigade at Schofield, having been almost killed in Iraq by an IED. But he still remembered Bobby's thoughtfulness.

Next came Air Assault School, which this time Bobby passed. The second time around he gave it the intense concentration and careful preparation it required.

Just before graduation, Bobby had a conversation with his father that illuminated the quality of military training Bobby had received at West Point. He had asked his father for advice on how best to deal with the power dynamics of the lieutenant-NCO relationship so central to the effective functioning of an infantry platoon: the lieutenant has the authority, but the NCO has vastly more military experience. As the conversation went on, Bob Sickler said to his son, "Do you have any idea how much more prepared you are to lead a platoon than I was when I was getting ready to be a platoon leader?" His point was that the USMA's military training had taught Bobby to think through the challenges involved in the lieutenant-NCO relationship before taking command in a way the Marine Corps Officer Candidate School had never done for his father during the Vietnam War. The challenges that West Point had taught Bobby to deal with before he was commissioned were challenges that his father only discovered a year into platoon command. Coming from his father, whom Bobby respected and whose advice he valued, this insight built Bobby's confidence in his ability to lead and manage a platoon.

Graduation day was one of the defining moments in Bobby's life not only because he felt a quiet sense of accomplishment but also because his father, a retired Marine colonel, accepted him as a brother officer. It happened while they were sitting on the deck at the 49er Lodge, a large log house in dense woodlands near Delafield Pond at West Point. Bobby and his friends were hosting a graduation party for their families. His father was a stoic who never expressed much emotion, never said "I love you, son," but now spoke to Bobby with a depth of feeling he had never expressed before. Bob was welcoming Bobby as a brother officer, a fellow member of an exclusive military fraternity, leaving no doubt how proud he was of his son.

Years later, Bobby reflected that West Point could not have done a better job preparing him to become a platoon commander. But it had also inspired him, "an average cadet" in his words, to learn to appreciate his full potential. The USMA had given him the priceless opportunity to learn that even though some goals were initially beyond his reach, with the right amount of drive and effort he could eventually achieve them. Bobby's father had sown the seeds. West Point provided the right environment in which they could grow and blossom.

11

STEPHEN TANGEN

R-Day

Stephen Tangen did not come from a military background and did not know what to expect at West Point on R-Day, June 28, 2004. At eighteen years of age, he was young, naive and understandably a little immature. Later he recalled, "You're at Eisenhower Hall, your parents have sixty seconds to say goodbye to you, you walk through a door, and that's the last time your parents see you. No phones, no internet, no nothing. Coming from high school to Beast Barracks not really knowing what I was getting myself into was a very rude awakening, and I started to struggle very quickly."

Stephen's first mistake was when he reported to a cadet in a red sash. He had no idea who she was but mistakenly introduced himself as new cadet Stephen Tangen. She replied by berating him. Why was he telling her his first name? Worse, up until that point Stephen had never had to do any rote memorization, so being handed the cadet "knowledge" book and told to memorize it was a serious challenge. He struggled with it throughout Beast Barracks.

Beast Barracks: Cadet Basic Training

Beast Barracks was and is six weeks of mental and physical purgatory designed to test the physical and mental resilience of the new Plebes. For Stephen, Beast Barracks became a traumatic experience, a turning point in his life. Physically fit, he had little difficulty with the intensive physical training program. The problem was that some members of the cadet cadre supervising his squad enjoyed abusing Plebes. Matters came to a head when a female USMAPS alumni in his squad refused to perform her push-ups. In response, the cadet cadre punished and hazed the rest of the squad. Stephen was surprised and shocked. He had read a lot about West Point and expected Beast Barracks to be demanding. But he never thought he would be demeaned like this. At home as a boy, Stephen had been occasionally punished for bad behavior, which he accepted. But his parents had never punished him for his brothers' bad behavior.

Stephen and his squad mates thought that the USMAPS alumna had refused to do either her physical training exercises or corrective training because she knew that under Cadet Basic Training rules, the cadet cadre could not touch her. So, if she just stood and cried, she would not have to do anything she didn't want to. Now, no one on the cadet cadre explained why they were making Stephen and his squad mates do the other Plebe's push-ups. But he and his squad mates thought that maybe they were doing this to make them help her perform the training tasks that she was refusing to do. The squad tried but failed: she stood by and watched the rest of the squad get punished on her behalf. As Stephen saw it, she did not seem to care. And so, the cadet cadre continued to verbally abuse and haze the squad. They even gave them "extra duties." These included helping her to wear her uniform correctly and reciting cadet knowledge on her behalf. Worse, Stephen and his Beast Barracks roommate, Mike Roth, were sometimes singled out for additional abuse because they were Corps Squad athletes. The cadet cadre apparently had a grudge against Corps Squad athletes because they are exempted from some formations, drill, and ceremonies in order to train, travel, and compete for the USMA.

Understandably, Stephen was angry: he had arrived at West Point as an All-American scholar-athlete and, understandably, thought he deserved better treatment. His message to the cadet cadre was, "Why are you expecting me to do someone else's punishment? And why do you expect me to be okay with that?" He told his squad leader that he wanted to speak to the cadet counselor (usually a third-year cadet). Stephen told the cadet counselor that he was being punished because another member of his squad refused to perform, and he wanted to know why. He added that he did not know how to persuade this USMAPS alumna to perform, much less how to get out of what he saw as a trap. The cadet counselor did not know what to say, dismissing the whole problem with a casual "Oh, that's just the way it is. You are just a quitter." Stephen felt "broken." Over the Fourth of July holiday weekend, he called his father for advice.

For Stephen, the turning point was his call with his parents. Susan, his mother, mostly cried but tried to comfort him. Andrew, his father, offered him an uplifting story with similar beginnings but a positive outcome. His unscrupulous Air Force recruiter had lied to him. He had told Andrew that there was a way he could become a pilot. There wasn't. Instead, Andrew became a mechanic crawling inside the fuel tanks of WC-121 Warning Stars hurricane-hunters without a mask or any other protective gear. He also endured verbal abuse from his Air Force superiors. Andrew found solace caring for a stray kitten that lived beneath his barracks. He brought the kitten food and water from the mess hall so it would not starve to death. Caring for the kitten gave Andrew a chance to escape from the betrayal and disappointment of his Air Force enlistment. Throughout that enlistment, Andrew faithfully followed orders and carried out his duties. During the phone call Andrew told his son that "I know where you are. I've been there. I know you are struggling, but there's something more important to keep focused on. You can get through this." Andrew's message to his son was that he had endured a parallel hardship at a similar young age and had overcome it. Although

Stephen did not have the solace of a malnourished kitten to care for, he decided that he would not break. He would overcome the unfair punishment he was receiving and would continue to receive. He would become the best cadet he could possibly be, even cadet first captain. In that role, he could run the entire corps. He didn't make first captain, but by his Firstie year he had risen from nine hundredth on the military ranking list to sixteenth. It was a stunning achievement.

Before the end of his Beast Barracks ordeal, however, worse was to come. At the end of the six weeks, Stephen reported to his TAC officer, who opened the conversation by saying, "I am seriously worried about you. You have essentially failed Cadet Basic Training. Are you really with us? Do you want to go home?" Stephen was shattered. His one goal in life had been to go to West Point and excel. Instead, although he did nothing wrong, his TAC officer was now telling him that he was failing. Instead of trying to solve a legitimate problem, the TAC officer was as dismissive as the cadet counselor. Later, Stephen thought that the only thing the TAC officer was really worried about was whether he might attempt to commit suicide by jumping from his sixth-floor room in MacArthur Barracks. That was *never* going to happen, but the TAC officer had completely misread Stephen because he had never taken the time to understand him, much less the original problem in the squad.

During Beast Barracks, Stephen really missed his family. Like many older teenagers, he had not always been kind to his parents during his junior and senior years in high school. And once he was accepted to West Point, his attitude toward them was "I don't need you guys anymore. You aren't paying for my college. I am going to do this. Let me do whatever I want to do." But once his parents were no longer present, once he lost what he once had, once he had to rely on his father to help him through a crisis, he realized how important being with family really was.

In late August once the academic year began, however, some things got better. Stephen enjoyed his academic courses; swimming began too, so he was back in his niche in the pool on the USMA's varsity swim team. And at last he had some good luck: he was assigned a great roommate, Jim Villanueva, a determined, academically gifted young man from Bloomfield, New Jersey. Although Jim had not experienced any abuse at Cadet Basic Training, he too was something of an independent thinker, an outsider experiencing difficulties adjusting to West Point's exceptionally demanding environment. Jim, for example, wanted to focus more on his academic courses and less on his Plebe duties, something his cadet chain of command didn't like. Like Stephen, Jim had no strong military tradition in the family and, although he had visited West Point, he was finding life as a Plebe far more difficult than he had expected. Jim and Stephen were both determined not to fail and not to quit. So, together they formed a pact; as Jim later described it, "Iron sharpens iron." Every evening, they talked through their academic and military assignments, how to manage time, what to focus on, and how best to approach it. Stephen helped Jim improve his swimming where he struggled.

Every morning, Jim and Stephen also read the *New York Times* with a special focus on the Iraq War. By the fall of 2004, the war was not going well. The two cadets discussed how the war was being fought, what was going wrong, and why.

The USMA assigned Stephen and Jim new roommates for the spring of 2005, but they continued to support each other, becoming lifelong friends. Looking back, Stephen said that "without Jim, I don't know if I would have made it all the way through West Point."

This was especially true in that first semester, when members of the cadet cadre who had supervised him during Beast Barracks regularly taunted him as a "quitter" and a "dirtbag." Even his new cadet chain of command verbally abused him, trying to pressure him to quit West Point. Every day they taunted him, saying "we're going to get you to break." Despite this torrent of taunts and verbal abuse, Stephen acquitted himself, earning a respectable 3.2 GPA by the end of the fall semester in December 2004. He fell below a B only in his military development course and in boxing.

From January 2005 onward, Stephen had a different chain of command. There was no more verbal abuse, no more hazing from his cadet cadre. His new chain of command could see that Stephen cared about

West Point, that he was eager to learn, grow, and do everything he could to become cadet first captain. His academic grades improved still further, and his military development and military science grades soared. He now began his dramatic rise to sixteenth in his class military graduation ranking. Absent the unfairly low grade in Cadet Basic Training, he would have been one of the top three cadets militarily in his class.

Plebe year was challenging, but Stephen understood and embraced the new leadership model put in place by then commandant of cadets, Brig. Gen. John Abizaid, back in 1995. However, Stephen thought that "most of the traditional stuff was not changed by the new leader development philosophy: the Plebes continued to take out the trash, deliver the newspapers and the dry-cleaned uniforms." Still, Stephen understood and embraced the new West Point philosophy: becoming a more empathetic leader by first learning to follow. Plebes at West Point carried out the same duties as soldiers in the field army. So, your Plebe experience would help you understand how your soldiers feel and think.

From January 2005 onward, Stephen's military science classes took on a new importance. As the United States entered the fourth year of conflict in Afghanistan and the second year in Iraq, the Army was learning new lessons and developing new doctrines and new ways of organizing to fight, and these were being quickly incorporated into the military science curriculum at West Point. As Stephen recalled, "West Point's Military Science Department did an extraordinary job. They were really on it." The Army was reorganizing its structures to fight, as well as developing new ways of waging war against insurgents in Iraq and Afghanistan. All Stephen's instructors were combat veterans, and in his eyes this gave them a special credibility. He listened and learned eagerly. He earned excellent grades because of the way he was able to grasp new concepts and new doctrine taught by these combat veterans. Stephen was learning how to function as an officer in wars unlike any the United States had fought before.

In the summer of 2005, Stephen began his eight-week Cadet Field Training program. The first four weeks were very individually focused:

rifle marksmanship, rappelling, and reconnaissance. The second four weeks were all team-focused on squad- and platoon-level exercises, patrol-based operations, and learning how to operate from forward operating bases (FOBs). The new concepts, new doctrine, and new ways that the Army was being organized to fight had been taught in the military science classes since January 2005 but unfortunately were not integrated into the Cadet Field Training exercises until the summer of 2006.

In the fall of 2005, Stephen began his second, or Yearling, year. At this point in his West Point career he took responsibility first for one and then two plebes, part of the Abizaid leadership model. Stephen welcomed the new responsibility, but as he later recalled, "It's kind of a taste of responsibility for someone else, which depending on the luck of the draw, depending on which Plebe you get, it can be a very rewarding experience or a miserable experience." Stephen added,

I had one Plebe my first semester and two my second semester as a Yearling. The first semester was a struggle. This Plebe was a very bright young woman. Academically she did very well but really did not enjoy the West Point culture. She ended up leaving West Point at the end of her Plebe year. In the second semester, I had two young men. One was very strong, one very weak. The leadership challenge there was how to balance the two. Do you adopt the sharper tone? Do you make the better one responsible for the weaker one? Or do you invest different amounts of energy to both, or do you mix the strategy?

Stephen took a balanced approach and spent coequal time with both individually, but he strongly encouraged, coached, and mentored them to help each other as a buddy team.

For Stephen, an even bigger challenge was finding the time to mentor these Plebes in the way he wanted to. During the academic year, Stephen's days were jam-packed. Like all cadets, he was taking five courses and was also on the USMA swim team with a tough, time-consuming training schedule. Stephen did everything he could for the two Plebes he was responsible for, but the severe pressures on his time meant that in reality there were very few opportunities to guide them.

In the summer of 2006, Stephen returned to Beast Barracks, now as a member of the cadet leadership cadre. Because of his own bad Beast Barracks experience, he had volunteered to serve as a cadet counselor. This involved a five-week Beast Barracks commitment, not the usual three weeks. Serving as a cadet counselor not only guaranteed him airborne school but also an extended eight-week vacation the following summer when he only had his CTLT to complete. That meant eight weeks with his beloved girlfriend, Melissa, and his family.

The pressures on Stephen's time were made more intense by his determination to work on his military performance. After overcoming the problems during Cadet Basic Training, he was determined to seek out opportunities for cadet leadership within the USMA. After a difficult Plebe year, Stephen's performance had improved dramatically. By the fall of 2006, the beginning of his Cow year, Stephen became company first sergeant. In the spring semester, he worked hard for promotion to battalion sergeant major but didn't succeed. Undaunted, he took on the additional duty of company administration officer. Making Excel spreadsheets for the company was thankless work, but it gave him the opportunity to represent the company and go before the Emergent Leader Board, where he could compete to be one of the regimental commanders for summer training.

Unfortunately for Stephen, the Emergent Leader Board was a "murder board" in which the odds were against him because he could not possibly know all the answers. Led by an Army sergeant major, the questioning was rigorous, relentless, and tricky. The question that stumped Stephen was "Cadet Tangen, you are walking to class and you see a uniform deficiency. What do you do?" Stephen replied, "Sergeant Major, I would stop and correct the deficiency." The sergeant major asked the question three times, but Stephen didn't see what he was looking for. It was a trick question because if you stop and correct every uniform deficiency you see, then you won't get to class on time and will be punished by having to walk for hours, in full dress uniform carrying a rifle, in the "Area" irrespective of how bad the weather was. The answer the sergeant major was looking for was "I wouldn't stop because I cannot personally correct everything that is wrong." Not answering correctly caused Stephen to fail his Emergent Leader Board review. He had no regrets, however. He had stuck to his core values: if he saw a blatant violation, he would fix it and told the sergeant major that three times.

In the summer of 2006 between his Cow and Firstie years, Stephen had to do his CTLT. This is the Army's version of a summer internship during which cadets are matched with lieutenants in the field army at any one of sixteen U.S. Army bases in the United States and overseas bases in Germany, Italy, South Korea, and the country of Georgia. The list changes annually. Cadets are not, however, placed in combat zones. The goal is to help cadets understand all aspects of how a company works as well as the role of a new lieutenant in it. It's a great learning opportunity.

Cadets get to rank their choice of branch and their five preferred locations. Stephen applied to go to Germany because he had studied German in high school. Unfortunately, on the Friday before final exams, his TAC officer called him to say "Sorry, you are not going to Germany." Stephen didn't go because a Filipino cadet selected to be regimental S4 quartermaster for Camp Buckner summer training had already bought a $2,000 air ticket to the Philippines to visit his family. Because he was an allied cadet, the USMA decided it did not want him to miss his family visit. And so, Stephen was told he would become the Filipino cadet's replacement as regimental S4 at Camp Buckner but later on could have whatever CTLT he wanted. "What CTLT assignments are still available?" Stephen asked. His TAC officer replied that Hawaii was open, and Stephen accepted that assignment. By Monday, however, Hawaii, was not available. Now the last CTLT available was South Korea, which Stephen accepted.

Stephen was bitterly disappointed. The Army had scrapped his plan to spend eight weeks with Melissa, something that caused her to temporarily break off their relationship. Stephen was not happy about the Cadet Basic Training assignment, but the regimental logistical job there turned out to be an unexpectedly good opportunity. Indeed, at the end of the summer the civilian cadet corps S4 gave Stephen one of the nicest compliments he had received at West Point to date: "I've been here for over twenty years, and you are the best logistics officer that ever worked with me." Stephen's outstanding performance earned him a regimental staff position, S3, in the fall semester and regimental XO in the spring. It was not first captain as he had always hoped for, but it was recognition of how far he had climbed in his military ranking as well as an opportunity to continue to develop his leadership skills.

After all the difficulties in getting his CTLT assignment sorted out, Stephen's South Korean assignment turned out to be fun. When Stephen arrived, there was no formal CTLT assignment for him. Indeed, his hosts had not even been expecting him. So, Stephen ended up being assigned to a group of U.S. foreign area officers who escorted him through a grand tour of South Korea.

In the fall of 2007 Stephen renewed his working relationship with Nick Eslinger, who was now regimental sergeant major. During Stephen's Plebe year, Nick had been his first sergeant at Camp Buckner. Stephen was much impressed with Nick's leadership. Watching his example, Stephen concluded "that's the kind of leader I would like to be." Little did he know that in 2009, he would take over Nick's platoon when it returned from Iraq.

In the fall of 2007, the beginning of his Firstie year, Stephen was also reunited with Jim Villanueva when once again they became roommates. Every morning they read the *New York Times* together, focusing once again on the Iraq War that by then had bogged down in a stalemate. As the year went on, it was clear to Jim that Stephen had grown into a strong young leader who could inspire and motivate. Stephen was now more focused, more confident in leading Plebes and other junior classmen. He was justifiably proud of his academic success in the life sciences, one of the USMA's most demanding majors, earning a 3.6 cumulative GPA. He was also justifiably proud of his military success rising from nine hundredth at the beginning of Plebe year to sixteenth by graduation.

What lay behind Stephen's remarkable success at West Point? There were three coequal reasons. The first was Stephen's own inner resilience and stubborn determination to prove his early detractors wrong. As his roommate Jim said, "Stephen feeds off people telling him he can't do it. He will do everything he can to prove them wrong." And he did. Everyone who had told him that he wasn't good enough to succeed at West Point were proven wrong. The only thing they accomplished was to build Stephen's fierce determination to succeed and prove them wrong.

The second reason was moral: Stephen's passion to attend West Point came directly from his personal struggle to become the best person he could be. He had been seeking a purpose and direction and had found it at the USMA, which upheld the same moral values he had held throughout his life, and he was determined to become a successful military and moral leader.

The third reason is that, like his best friend Jim Villanueva, Stephen was determined to discover his full potential as a young man and as an officer. Together, they decided that the best way to do this was by pushing their limits, trying to find out how well they could do. And they did.

Stephen graduated from West Point in May 2008 with a BSc in chemistry. He was now a strong, confident, focused leader well prepared for platoon-level command. In overcoming adversity at the USMA, he built an exceptional level of resilience as well as empathy for young soldiers. Militarily, Stephen had embraced the Army's new concepts and doctrines with enthusiasm, never earning anything less than an A– in his military science classes. Academically, his performance was also outstanding across a wide range of disciplines, with stellar grades in demanding chemistry courses. Despite all of the adversity he faced, Stephen never failed to earn a place on the dean's list.

PART 3

PREPARATION

12

RANGER SCHOOL

Ranger School is purgatory. It is one of the most elite Army training schools, often described as "the toughest combat school in the world" or, in Ross Pixler's words, "Beast on steroids." Ranger School is based at Fort Benning, Georgia; Camp Merrill, Georgia; and Camp Rudder, Eglin Air Force Base, Florida. As Maj. John Spencer writes, "Ranger School was developed in 1951 during the Korean War after Ranger companies, made up of volunteers who underwent intensive specialized training, demonstrated overwhelming combat proficiency on the Korean battlefield."[1] Korean and World War II veterans "emphasized the importance of individual combat skills, mental and physical toughness, and decision-making under extreme stress—skills that remain the focus in Ranger School today."

Any member of the active-duty U.S. military, male or female, can apply to Ranger School; you do not have to be an officer. The upper age limit is forty-one, although the overwhelming majority of Ranger students are in their twenties. But you must be mentally and physically fit.

The goal of Ranger School is to teach young officers how to lead soldiers through platoon-level missions in wooded, mountainous, and swamp conditions in the most stressful situations imaginable when they themselves are sleep-deprived, exhausted, and hungry. Ranger candidates endure twenty hours of operations a day fueled by at most two MREs (meals ready to eat) per day. Ranger instructors use a carefully calibrated scientific method to determine the level of physical and psychological stress to subject each Ranger class to. The goal is to subject each Ranger candidate to the maximum possible physical and psychological stress without causing serious injury or death. It's brutal.

For any young infantry officer hoping to make his mark, this is the school you have to graduate from if you want to have credibility with your chain of command and with your soldiers. It is a rite of passage for infantry officers. When you arrive at your first unit, everybody checks your uniform's upper sleeve to check that the Ranger tab is there. If it's not, you have started your career on the back foot.

Throughout Ranger School you have to choose your pain: sleep deprivation or hunger. Ranger instructors give you only two to four hours of sleep after a mission that might have ended at 1 a.m. You are woken up at 5 a.m. But within those four hours you have to eat and pull security for your squad for thirty minutes. So, in reality you only get two and a half to three hours of sleep. The impact on your cognitive ability is significant. At the end of multiple days without enough sleep or food coupled with constant physical activity, you start to "drone": you walk like a zombie, and your decision making is severely degraded: Can you cope? Can you still lead patrols and carry out missions effectively in this condition? Can you make the right decisions?

The first phase, Darby, begins at Camp Rogers at Fort Benning. It begins with a tough round of physical fitness tests before moving on to basic soldiering skills, basic tactical competence, and squad-level leadership. The Darby phase poses a series of questions for Ranger candidates: Can you do "49 push-ups, 59 sit-ups, a five-mile release run finished in 40 minutes or less and six chin-ups?"[2] Can you also pass the combat water survival test, do the Darby mile run, complete more five-mile runs, navigate the notoriously difficult Darby obstacle course, and tackle a twelve-mile march carrying over sixty pounds of equipment? Each day in the first week of Camp Derby there is an assessment that you have to pass. If you fail, you have to retest. If you fail the retest, you are out of Ranger School.

Darby also requires that you successfully execute a parachute jump from a fixed-wing aircraft, land navigation exercises in daylight and darkness without a GPS, and master the basics of planning and executing patrols and small-unit tactics. Here, you are graded on the basis of your performance in Ranger instructor-led and candidate-led tactical operations.

If you pass the Darby phase of Ranger School (over half of candidates don't), then you move on to Camp Merrill in Dahlonega in the foothills of the Blue Ridge Mountains. Founded in 1952, Camp Merrill is named in honor of Maj. Gen. Frank Merrill, commander of the famous Merrill's Marauders commando unit in World War II. If you arrive in winter, as Nick Eslinger, Ross Pixler, and Stephen Tangen did, it is a cold, wet, often snowy place. Average temperatures are a few degrees above freezing during the day and a few degrees below at night. Average rainfall is three times that for the rest of the United States. In the winter, there is heavy snowfall.

At Camp Merrill, the emphasis shifts to platoon-level leadership in mountain warfare. Ranger candidates have to learn how to prepare platoon operation orders and lead platoon-level missions that involve rock climbing, rappelling, and lowering equipment down ridgelines. Ranger missions take place in the rugged 750,145-acre Chattahoochee National Forest.

The final or swamp phase of Ranger School takes place at Camp Rudder next to Eglin Air Force base in northern Florida. As the official Army website describes it, the goal at Camp Rudder is to ensure that Rangers can "develop the student's ability to plan and lead small units during independent and coordinated airborne, air assault, small boat, and dismounted patrol operations in a combat environment against a determined and well-equipped . . . opposing force."[3] In all, there are two missions in the swamps and a simulated commando raid on military facilities on Santa Rosa island off the coast of Pensacola, Florida. In this final stage, Ranger instructors (RIs) step back and give Ranger students more room to devise imaginative solutions to the more complicated and often unexpected problems they have presented them with.

It is often suggested that the key to success at Ranger School is the support of your squad mates, that at its heart, Ranger School is collaborative. That is largely true. As Nick Eslinger opined, "Individuals do not succeed at Ranger School; teams do." The only way to succeed is to work together as an integrated team, with each member taking responsibility for the success of the squad as a whole. When one of your squad does not get a go-ahead to the next phase of Ranger School or if you are leading the squad in an evaluated exercise and you don't pass, the whole squad feels like it has failed.

Building those integrated teams is not easy. The first reason is that the attrition rate is so high: people keep disappearing. The person you establish a relationship with today may not be there tomorrow. In the first week at Camp Darby, for example, about 50 percent of Ranger candidates are eliminated from the program. The second reason is that you get reassigned to a new squad during Darby, Merrill, or Rudder or to a new platoon after each phase of Ranger School. A third reason is that you get squads that just don't like each other, and this can be reflected in unfair or unreasonable peer reviews. It's complicated. For any Ranger candidate, the key to success is being a selfless servant-leader who has a reputation as a contributor: someone who does everything they possibly can to help classmates, the squad, or the platoon succeed.

13

NICK ESLINGER

On May 26, 2007, Nick Eslinger graduated from West Point with a bachelor's degree in military science. He chose the sixty-day leave option and returned home to his parents. By this time, Donna and Bruce Behnke had moved from California to Houston, Texas. Nick spent the summer of 2007 decompressing from the rigors of his West Point experience in the hot Texas sun and reconnecting with his beloved parents and younger half brother, Danny. Nick's next career step was the Infantry Basic Officer Leader Course (IBOLC) at Fort Benning, Georgia, in September.

In the U.S. Army, physical fitness is always a key litmus test. Nick knew that if he was to make a good first impression on his instructors at Fort Benning, he needed to be as fit as possible. To that end, he undertook weekly road marches, five gym workouts per week, and disciplined nutrition throughout the summer.

Fort Benning, "home of the infantry," is a massive 182,000-acre base. It is a major training and development center for the U.S. Army, housing numerous training schools ranging from the NCO Academy to Army Ranger School, the toughest leadership training school in the world. When Nick arrived in September 2007 at the height of the wars in Iraq and Afghanistan, Fort Benning's training schools were operating at full capacity. For Nick, the IBOLC was extremely beneficial. At West Point, he had learned not only to be a principled leader of character but also the basic technical and tactical competencies necessary to succeed at Fort Benning. There, he learned his craft as a junior infantry officer: field training, field tactics, fire planning, how to navigate terrain, how to move and deploy his platoon effectively, and how to communicate infantry doctrine effectively to his soldiers as well as how to plan, implement, and evaluate his platoon's training for combat.

Taught by experienced commanders fresh from the battlefields of Iraq and Afghanistan, Nick's classes were infused with hard-earned wisdom. Determined to be the best platoon leader he could be, Nick absorbed all the valuable lessons these battle-hardened veterans had to teach. He graduated near the top of his class, well equipped physically and tactically for Ranger School. Everything his class needed it got. And on a personal level, Fort Benning was a really enjoyable experience because Nick was able to share a house with some of his close friends from West Point.

Next up was the purgatory of Ranger School. Throughout Ranger School you have to choose your pain: sleep deprivation or hunger. Nick chose sleep deprivation because he found dealing with hunger more difficult. Ranger instructors give you only four hours of sleep after a mission that might have ended at 1 a.m. Wake-up is at 5 a.m. But within those four hours you have to eat and pull security for your squad for thirty minutes. So, in reality you only get two and a half to three hours of sleep. The impact on your cognitive ability is significant.

The fact that hunger was harder to cope with than sleep deprivation for Nick, he always chose to eat. Although this bolstered his strength, sleep deprivation took a toll. To begin with, it caused him to become agitated very quickly. Little things that his classmates did that would not otherwise have bothered him now did. But Nick saw this as an opportunity to become more self-aware and did all he could to regulate his own behavior.

Sleep deprivation also took a toll on Stephen's decision making. His ability to absorb information and to make a quality tactical decision quickly deteriorated. But instead of making hasty unilateral decisions, Nick did the opposite: he got a second opinion, got new ideas and fresh perspectives, and only then made his decision. He drew the right lesson: in combat where leaders are often sleep-deprived, decision making should be cooperative.

As a new lieutenant, Nick graduated from the IBOLC course equipped physically and tactically for Ranger School. He was highly motivated to succeed not only because he wanted to meet everyone's expectations but also because he wanted to prove to himself that he could pass the best leadership school in the U.S. Army.

For Nick, the most difficult challenge at Camp Darby was not executing the physical training tests, the five-mile run, the twelve-mile march, the obstacle course, or the land navigation exercise. Rather, it was overcoming the fear of failure. In the minds of young officers such as Nick, Ranger School becomes "an absolute must" to graduate from. There was a lot of stress associated with the fear of failure at something that was absolutely expected of him.

In late February 2008, Nick arrived at Camp Merrill. It was a very cold place, with average temperatures in the day around freezing and often ten degrees below at night. At Camp Merrill, the emphasis is on platoon-level leadership in mountain warfare. As a prepster and as a cadet at West Point, Nick had some mountaineering experience in the USMA's annual Sandhurst competition against Britain's Royal Military Academy and other schools. And as a young man with steely resolve, he had the grit necessary to climb any Georgia mountain.

When Nick arrived at Camp Merrill in late February 2008, it was a bitterly cold day. His body was worn down from lack of sleep and food. Worse, the Ranger instructors did not allow Nick and his fellow junior officers to wear warm winter clothing, part of Ranger School's program to foster mental toughness. As a young man who was born and raised in the warmth of California, just dealing with the elements was Nick's greatest challenge, not only personally but also as a leader. It was so much harder to motivate his squad or platoon to accomplish the mission when they were freezing cold as well. This was when Nick was forced to turn inward and ask himself "can I do this?" It was so cold. Nick was so tired and so hungry, and Ranger School was not even half over. It was in the middle of the Merrill phase that Nick wondered whether he was going to quit or not.

Every one of the young officers in Nick's class experienced this moment of doubt, but for him it came during the Merrill phase, the toughest part of Ranger School. Throughout the two five-day field exercises, he had to dig deep inside his reserve of inner strength. He told himself "this is what I am supposed to be feeling, and how I respond will say a lot about me. In making me tired, hungry, uncomfortable, and stressed, this school is making me a better leader. In combat I will likely face these circumstances."

In overcoming his doubts about his capacity to succeed in Ranger School, Nick also had the support of his squad mates, who helped him out of what he described as his "emotional hole," and he did the same for them. As Nick recalled, "You have to ask for help and have teammates ready and willing to help; individuals do not succeed at Ranger School, teams do." The only way to succeed is to work together as an integrated team where each member takes responsibility for the success of the squad as a whole. When one of your squad mates does not get a go-ahead to the next phase of Ranger School or if you are leading the squad in an evaluated exercise and you don't pass, the whole squad feels like it has failed.

This does not mean that all Ranger cadet relations were harmonious. They weren't. Nick observed friction between squads and even some Ranger students trying to game the grading system by negotiating good reviews from their peers.

The Ranger School policy of constant rotations—regularly moving Ranger candidates to new squads or new platoons rather than leaving them in the same unit—added another layer of difficulty to building this camaraderie. But for Nick this added level of difficulty helped him learn more about himself and his ability to lead. He saw that the key to success was to build a reputation as a good teammate, a good contributor. As Nick recalled, "What is important is proving yourself a good contributor, acting as a selfless servant; that's how you form a team and build respect around you." Thanks to his own inner strength and the support of his squad mates, Nick got a go-ahead to the final phase of Ranger School at Camp Rudder, Eglin Air Force Base, Florida.

After parachuting into Camp Rudder in mid-March 2008, Nick got what he described as a "burst of motivation" not only because he knew it would be warmer but also because he could finally see the finish line in sight. For Nick, the greatest challenge at Camp Rudder was staying focused on the task immediately at hand rather than thinking too far ahead. At this final stage of Ranger School, losing focus causes you to make mistakes. Nick didn't.

At this point, RIs from the 6th Ranger Training Battalion expect more of the leader of each exercise and judge candidates' field performance more rigorously than in either of the earlier two phases. The task for the young Ranger candidates is to apply all of the leadership skills and tactical and technical knowledge they learned at Derby and Merrill to exercises simulating raids, ambushes, and infantry assaults in rivers, swamps, and on Santa Rosa Island, a long, narrow barrier island just off the coast of Pensacola in the Florida panhandle. Inevitably, the grading at Rudder is tougher.

By this final phase of Ranger School, Nick and his squad and platoon mates had built real unit cohesion. That cohesion, that willingness to work together, to help each other over the finish line, ensured that the vast majority passed the final phase of Ranger School. Much of that cohesion was attributable to the respect and reverence Nick had earned in the first six weeks. Because of that, neither his squad nor his platoon was ever going to quit on him.

Nick's last mission at Camp Rudder was a World War II–style commando raid attacking installations on Santa Rosa Island. The Santa Rosa Island raid taught Nick some valuable new skills about amphibious warfare. Less than a year later in Iraq, he put them to good use when he executed a similar mission to an island and cleared it of insurgents.

Graduating Ranger School was one of the most fulfilling experiences of Nick's life. He had been pushed to the outer limits of his endurance but had conquered every obstacle, including himself. After Ranger School, Nick had felt fully prepared to lead tough, physically demanding patrols against a well-trained, sophisticated enemy. "You learn so much about yourself at Ranger School, and you come out with confidence," he said. "I learned and became more self-aware of the point at which I start to drone: how far I can go without water, food, sleep, or rest before I am incapable of making good decisions. I also learned that there actually isn't a point where I could not do something. I can do it. I proved it to myself at Ranger School."

After Ranger School, Nick had a fresh appreciation for the simple joys of life: a hot meal or a hot shower when you want it. He especially enjoyed being able to control his own physical environment again. Driving from Fort Benning to his parents' home in Houston, Nick loved being able to adjust the climate controls in his truck. If he was hot, he could cool down. If he was cold, he could warm up. After Ranger School, this was luxurious.

14

ROSS PIXLER

West Point's Most Holy Trinity Catholic Chapel is a gray granite late Victorian church with a soaring white spire. It sits atop a hill just off Washington Road overlooking the Hudson River. There, on May 26, 2005, Ross and April Pixler were married. He had graduated West Point the day before.

Ross never expected to do that. Indeed, he used to make fun of cadets who got married at the USMA the day after graduation because he thought it "so clichéd." But in the end practicality prevailed. The only day that Ross's widely scattered extended family could come together with April's New York–based family for the wedding was at his West Point graduation. And it all worked wonderfully. Ross's older brother, Ryan, who had just returned from an overseas deployment, was best man. Afterward, the newlyweds headed off on a five-week honeymoon that included stays in Fort Lauderdale and Orlando in Florida, Las Vegas in Nevada, and the Pacific coast of Mexico as well as a Bahamas cruise and a Hawaiian cruise. To pay for it, Ross had moonlighted on weekends working in April's uncle's tree service business while a cadet at West Point. To reduce the cost of the honeymoon, Ross researched every possible every travel bargain.

By the time Ross reported to Fort Benning in late July 2005, he and April were almost broke. They had spent their money on the honeymoon, and there was little left with which to buy a house. So, they bought a dilapidated fixer-upper off post across the river in Alabama. An old blown-out television held the front door in place, and all the floors and walls had to be removed and replaced. Everything had to be gutted. Over the next four years, Ross and April had to refurbish every room in the house.

Fort Benning, "home of the infantry," is a massive 182,000-acre base. It is a major training and development center for the U.S. Army, housing numerous training schools ranging from the NCO Academy to Army Ranger School, the toughest leadership training school in the world. When Ross arrived in late July 2005 at the height of the wars in Iraq and Afghanistan, Fort Benning's training schools were operating at full capacity.

Here Ross began his formal army training. First up was Airborne School, a tough three-week training program in which officers and enlisted personnel learn how to jump out of an aircraft and how to land safely. It culminates in jump week with five jumps from a fixed-wing aircraft flying at 1,250 feet, three with full combat equipment and two without. Ever the adventure seeker, Ross was in his element and completed the course as the honor graduate of his class.

Next came the IBOLC. Ross thrived in this setting. West Point had taught him to be a principled leader of character and also the basic technical and tactical competencies necessary to succeed at Fort Benning. There, he learned his craft as a junior infantry officer: field training, field tactics, fire planning, how to navigate terrain, how to move and deploy his platoon effectively, how to communicate infantry doctrine effectively to his soldiers, and how to plan, implement, and evaluate his platoon's training for combat.

Taught by experienced commanders fresh from the battlefields of Iraq and Afghanistan, Ross's classes were infused with hard-earned wisdom. Determined to be the best platoon leader he could be, Ross absorbed all the valuable lessons these battle-hardened veterans had to teach. On December 17, 2005, he graduated from the IBOLC near the top of his class, well equipped physically and tactically for Ranger School.

Ranger School

After Ross graduated from the IBOLC he was unable to go to Ranger School because his branch manager had failed to secure any slots in the program. No one had aligned the start dates of the different courses at Fort Benning. Like all of his fellow West Pointers who had branched infantry, Ross had to wait. Soon he got fed up sitting around without anything to do, so he tried to get into different courses at Fort Benning to learn more and develop his skills. The only course that was open was a two-week second-level combative course. This involved learning to fight: hand-to-hand combat as well as offensive and defensive grappling. Ross described it as "fun" but conceded that "every one of us was getting beat up every day." He recalled that "my joints were crying when I graduated." Two days later, he began Ranger School. Ross's joints and muscles still ached; his elbows were hyperextended. It was a constant struggle to maintain good health, put 110 percent effort into every task he was given, and avoid getting hurt.

As a new lieutenant, Ross graduated from the IBOLC equipped physically and tactically for Ranger School. He was highly motivated to succeed not only because he wanted to meet everyone's expectations but also because he wanted to prove to himself that he could pass the toughest leadership school in the U.S. Army.

Ross's Ranger School experience was different than most. As we saw earlier, the branch representative for the IBOLC failed to secure any Ranger School slots for the three hundred students who has graduated from the course in November and December 2005. For the first three Ranger School cycles of 2006, only a handful of young infantry officers got in because those who had the slots had fallen ill or become injured before the school began. As a result, for three months there was a large backlog of young second lieutenants sitting around Fort

Benning waiting to go to Ranger School, the vast majority of them West Pointers.

To deal with this large backlog, the Ranger School decided to hold a special program to enable the three hundred young West Point graduates to earn their Ranger qualification. To do this, the Ranger School leadership had to call back every instructor who was on leave and cancel leave for those instructors about to depart. It was a necessary but unpopular decision.

Ranger School promised the young West Pointers that, despite the unusual circumstances, all the instructors would be professional and would act exactly as normal. According to Ross, that "did not exactly happen." He believes that there was a deliberate attempt to make his Ranger class's experience more difficult than it should have been. The statistics support Ross's concern: only 17 percent of his class graduated and earned their Ranger Tab, far below the average of 40–45 percent. Only 7–10 percent made it through without recycling, a number not in line with classes before or after. Ross himself was recycled after the first phase of Ranger School and had to start again.

The psychology of Ranger School is always difficult: trying to make sound military decisions under pressure while overcoming exhaustion, hunger, and sleeplessness is extremely taxing. For Ross, the psychological dimension of phase one of Ranger School was especially challenging. April was pregnant with their first child, and Ross was wracked with worry. He was not receiving her letters, and she was not receiving his. Not knowing whether or not his wife and unborn baby were doing well, not knowing if there were any medical problems, was deeply worrying. "I remember being very concerned," Ross later said. "I was going to be a father, and if anything went wrong, I couldn't be there." Understandably, he was somewhat distracted. Had the RIs told him that his mail was being held until he completed each phase of Ranger School, Ross's worries might have been alleviated to some degree. Ross was recycled and had to start again, but he did at least get his mail, including April's letters. Reassured that his pregnant wife was doing well, he began the first phase of Ranger School once again. This time he succeeded.

When Ross arrived at Camp Merrill, the second phase of Ranger School, in March 2006, it was a bitterly cold day. His body was worn down from lack of sleep and food; his muscles and joints ached. The RIs did not allow Ross and his fellow junior officers to wear warm winter clothing. This was when Ross was forced to turn inward and ask himself "can I do this?" It was so cold. Ross was so tired and so hungry, and Ranger School was not even half over.

During this difficult period, Ross often thought back to his experiences on Pico de Orizaba in Mexico and drew strength from it. He was cold in the first two phases of Ranger School but never as cold as he was on Pico de Orizaba. He was hungry in Ranger School, but he at least had something to eat every twenty-four to forty-eight hours. He may have been tired, but he never had to run for his life and that of his climbing partner over a full day. This helped put Ranger School in perspective, and Ross did not doubt his ability to make it through. He gained resilience of mind by tapping into his prior experience.

Ross loved hiking and mountaineering and so was looking forward to Camp Merrill. What he had not anticipated was the degree to which the RIs would deprive him and his fellow junior officers of sleep. In this second phase of Ranger School, you had to weigh the importance of sleeping or eating with special care. Ross tried to do both. RIs gave him only four hours of sleep, but within that period he also had to eat and pull security for his squad. So, Ross ate as quickly as possible, using that extra time to get a little more sleep. And he had one physiological advantage: once he lay down to rest, he was able to fall sleep almost immediately. But it was not enough. One night around 1 a.m., Ross hastily dug out a defensive position in which he could sleep. Exhausted, he fell asleep almost immediately. In the pitch-black darkness, however, he did not notice that he had dug himself into an ant hill. When he woke up, he found that the ants had bitten him all over his body. "It was horrible," Ross recalled.

More serious were the effects of chronic sleep deprivation. Ross recalls that one of the most vivid instances of sleepwalking during Ranger School occurred at Camp Merrill. He also experienced what he described as "really imaginative hallucinations." For Ross, as was the case for Nick Eslinger and Stephen Tangen, Camp Merrill presented the toughest challenge, the greatest moment of doubt. Throughout the entire nine-day field exercise Ross had to dig deep inside his reserve of inner strength.

The third and final stage of Ranger School took place at Camp Rudder, adjacent to the massive Eglin Air Force Base. As a native of Phoenix, Arizona, Ross especially enjoyed the warmer climate. The trials of Camp Rudder flew by fast for him. At that point, he had grown so accustomed to being in the field under severe physical and psychological pressure that going through the field exercises had become second nature. What was not second nature was the discovery of the body of a murder victim. Ross's Ranger buddy, Josh Silver, found a hand belonging to a murdered teenage girl when attempting to dig a slit trench used by Ranger students as a latrine. For Ross and Josh, the discovery was a profound shock.

It is often suggested that the key to success at Ranger School is the support of your squad mates, that at its heart Ranger School is collaborative. That is largely true. As Nick Eslinger recalled, "individuals do not succeed at Ranger School; teams do." The only way to succeed is to work together as an integrated team whereby each member takes responsibility for the success of the squad as a whole. When one of your squad mates does not get a go-ahead to the next phase of Ranger School or if you are leading the squad in an evaluated exercise and you don't pass, the whole squad feels like it has failed.

Here once again, Ross's Ranger School experience differed somewhat from that of the other elite souls. Ross found comradeship among his fellow West Pointers and learned that teams succeed at Ranger School, not individuals. But he also found some Ranger candidates who were not doing enough to help the team. Peer evaluations that are a part of Ranger School brought out some frictions. During these, Ross and others pointed out that some of their squad or platoon mates were not doing enough to help out their fellow mates. Reflecting later, Ross likened Ranger School peer evaluations to a game of *Survivor*, the TV reality show where people are voted off an island. He thought Ranger

School was similar in that at the end of each phase peers vote, and their votes can result in Ranger candidates being eliminated. And in Ross's view "there was a bit of politics in play" in this process in which factionalism within some squads and platoons played a part.

As Ross learned, preventing this factionalism can be difficult. And if you and your squad mates start off too friendly or too politely, it can end in disaster. Ross later recalled that in high-stress situations during Ranger exercises, "you can't be polite; you don't have the luxury of procedural niceties. If you want something done a particular way, you have to order it; you have to make sure that those around you are working hard to achieve the group's mission." This can, however, build resentment among fellow officers.

At the core of the final phase of Ranger School is an intensive ten-day field training exercise that takes place mainly in the camp's rivers and swamps. The task for the young officers is to apply their leadership skills and tactical and technical knowledge to exercises simulating raids, ambushes, and infantry assaults. Instructors from the 6th Ranger Training Battalion expect more of the leader of each exercise and judge candidates' field performance more rigorously than in either of the other two phases.

Traveling to a warmer climate, getting some much-needed sleep, and eating a hot meal on arrival at Camp Rudder reinvigorated Ross. By this final phase of Ranger School, he and his squad and platoon mates had built real unit cohesion. That cohesion, that willingness to work together to help each other over the finish line, ensured that the vast majority of those still remaining passed the final phase of Ranger School.

In June 2006, Ross graduated and earned the coveted Ranger Tab. For Ross, graduating from Ranger School was one of the most fulfilling experiences of his life. He had been pushed to the outer limits of his endurance but had conquered every obstacle, including himself. After Ranger School, Nick Eslinger thought he had felt fully prepared to lead tough, physically demanding patrols. "You learn so much about yourself at Ranger School, and you come out with confidence," he said. "I learned and became more self-aware of the point at which I start to drone: how far I can go without water, food, sleep, or rest before I am incapable of making good decisions. I also learned that there actually isn't a point where I could not do something. I can do it. I proved it to myself at Ranger School."

Ross agreed with Nick's assessment. In Ross's view, "Just because you are experiencing discomfort it doesn't mean you need to stop. Discomfort should not be allowed to make you want to stop. It don't mean you should stop or need to stop." Ross had first learned this lesson about endurance as a cadet in the Army's prescuba school. "You are holding your air," he said. "You think you are about to drown, you are feeling frantic, and then thirty seconds go by and you haven't passed out underwater. And yet thirty seconds ago you would have sworn that you were about to pass out. And then another thirty seconds go by, and you are still conscious. We are capable of so much more than we think we are. Once you get past that mental block, you identify how great your capability truly is."

15

STEPHEN TANGEN

Stephen Tangen graduated from West Point on May 31, 2008, with a bachelor's degree in chemistry. He chose the thirty-day leave option and returned home to Naperville, Illinois. He spent the first part of the summer of 2008 decompressing after four tough years at West Point with his beloved parents and extended family.

Professionally, Stephen's next step was the IBOLC, the course that trains second lieutenants to become infantry officers, at Fort Benning, Georgia, where he reported just after the Fourth of July holiday. His experience at Fort Benning was different than that of Nick Eslinger and Ross Pixler because Stephen first had to complete the eight-week Basic Officer Leaders Course. This was a new course the Army had introduced to give every one of its new officers, no matter which branch, a common eight-week training experience. As a West Point graduate, Stephen found the Basic Officer Leaders Course redundant with what he had already learned at the USMA: basic rifle marksmanship, familiarization with and firing other weapons, land navigation, convoy operations, and other basic soldiering tasks. The Pentagon canceled the Basic Officer Leaders Course the following year.

In early September 2008, Stephen started the IBOLC. He repeated the IBOLC phase on basic rifle marksmanship before moving on to advanced rifle marksmanship, which enabled him to learn useful lessons about close-order marksmanship, close-quarter battle, and clearing a house with live rounds. The third week brought land navigation: using the handheld Defense Advanced GPS Receiver, known as "dagger," to find locations anywhere day or night. In week four, Stephen participated in team-situational exercises in which he learned to maneuver soldiers as a team in both woodland and urban settings. Week five built on what he had learned with live-fire situational exercises from squad to platoon level: how to establish one side of a company perimeter and how to repel an attack and incorporate mortars.

For Stephen, IBOLC culminated in two weeks of live-fire platoon situational training exercises. It began with each officer rotating leadership of a platoon as well as filling different roles within it including squad leader and platoon sergeant. For almost all the time when Stephen was not leading a platoon, he was in its fourth heavy weapons squad as a machine gunner. IBOLC culminated in the larger Battle Force situational exercise involving multiple platoons. Here, Stephen learned how to lead a platoon as part of company-level missions.

On balance and with two important caveats, Stephen thought that the IBOLC had prepared him well for platoon leadership. It taught him all the minute details he needed to know: how to use all of the weapons, radios, and GPS equipment; how to call in supporting fire; and how to lead a squad and a platoon. For Stephen, the first important caveat was that he had only led a platoon five times at the IBOLC, each for a total of twenty-four hours. In his view, this was just enough time to figure out everything he is expected to do as a platoon leader and establish confidence in the role. Stephen would have preferred more time.

His second caveat was that although veterans of the wars in Iraq and Afghanistan who taught him during IBOLC brought invaluable experience and insight, there weren't enough of them. At that time each training platoon should have had two captains and six NCOs serving as guides and mentors. In practice, his IBOLC platoons only had one captain and three NCOs; some platoons only had one captain and one NCO. However, Stephen really appreciated the quality of the guidance his mentors offered. His principal mentor was a West Point graduate (class of 2004) who had deployed twice and was not only a constant presence but also a source of practical wisdom suffused with battlefield experience.

The IBOLC also prepared Stephen well for Ranger School. To pass IBOLC, he had to successfully complete the Ranger School "Darby Queen" obstacle course, its five-mile run, and its daily physical fitness test whenever he was not in the field. By the time he graduated from the IBOLC, he had passed Ranger School physical fitness tests a dozen times. His one criticism of IBOLC physical training was that it did not provide any weight-lifting equipment in the field at the time—today the IBOLC does. As a result, the only way to build up endurance and strength for Ranger School was by running and doing push-ups, pull-ups, and other body weight exercises.

Just before Thanksgiving 2008, Stephen passed the IBOLC. Once he returned to Fort Benning from the Thanksgiving holiday, he had two weeks before winter leave. During this time, all he had to do was report for morning physical training and for lunch. The rest of his day was free. But on that first Monday back, an NCO asked for volunteers for an unspecified task. Normally, this meant a mundane janitorial-level chore such as grounds maintenance. But Stephen thought that if he volunteered that Monday morning, he would not have to volunteer again before Christmas.

In fact, what he got was further help in preparing for Ranger School: two weeks of the Basic Combatives Level Two course. In Basic Combatives Level One, the focus is on learning how to maneuver your body and your assailant's body to gain dominance and avoid being grounded. The second level is more painful. It's about strike drills: defeating two people who are trying to hit you with punches to the thighs and shoulders. Unlike Ross Pixler, Stephen had winter leave to recover and allow his bruises to heal before he began Ranger School in the new year.

January 2009 was very cold at Fort Benning. Stephen remembers the daytime highs in the thirties with nighttime lows in the teens. For Stephen as for Ross and Nick, Ranger School was a grueling experience made worse by the cold.

As we have seen, Darby, the first phase of Ranger School, occurs on Fort Benning's sprawling campus. Darby begins with a tough round of physical fitness tests before moving on to basic soldiering skills, basic tactical competence, and squad-level leadership. Stephen was extraordinarily fit, but even he had difficulties with the water obstacle course on an early morning with the temperature twenty degrees below freezing. In the frigid waters littered with frozen clumps of ice, Stephen's leg muscles seized up. Despite the severe discomfort, he completed the course.

During the Darby phase, the Ranger NCOs devised a tough regimen for Stephen and his class. For ten days, they allowed them only two MREs per day and less than two hours of sleep per night. Moreover, they did not allow the Ranger candidates to eat while planning or executing or recovering from a mission. So, for the ten days of operations, there were twenty-two hours of activity and only two hours left to eat, sleep, and do mandatory guard duty. Stephen had a clear choice: sleep for ninety minutes and eat little, or eat two MREs and get sixty minutes of sleep. He chose to eat the two MREs. It was a wise choice. You can't prepare and eat two MREs in thirty minutes.

In addition to the demanding physical fitness exercises, Stephen and his class had to prove that they had mastered essential soldiering skills: assembling and disassembling weapons as well as executing a successful parachute jump from a fixed-wing aircraft. And then there was the important task of learning how to prepare and draft a platoon operations order that the RIs would approve as well as night navigation with only a map and compass. There was always a lot to do as Stephen coped with exhaustion and chronic sleep deprivation.

There was no time for rest or recovery between the end of the first Darby phase and the beginning of the second mountain phase at Camp Merrill. Stephen's final mission at Darby ended late in the evening in pitch-black darkness. Exhausted, Stephen carried his sixty-pound rucksack back to Ranger School headquarters for inspection by the RIs. He also had to produce his personal duffle bags for inspection and get ready to clean the barracks. The process began at 9 p.m. and was scheduled to last all night. The buses to take Stephen and his classmates to Camp Merrill were not scheduled to arrive until 5 a.m.

Like Ross, Stephen now had to endure what in this writer's opinion was unprofessional, abusive behavior. For several hours, Stephen and his classmates cleaned the barracks in preparation for a white glove inspection when everything would have to be spotless. It was, but an RI tasked with inspecting the barracks took chewing tobacco out of his mouth and threw it in the shower stalls. Screaming at the top of his voice, he accused the Ranger candidates of failing to clean the barracks to the required standard and of having contraband chewing tobacco. Both charges were untrue, but the lead RI now ordered Stephen and his classmates to do push-ups on the rock-strewn formation area until one of them confessed. Stephen was experiencing severe pain but kept doing push-ups. When no one confessed, the RIs ordered them to dump out their rucksacks and duffle bags on the ground. As they searched for the nonexistent contraband, the RIs threw the Ranger candidates' personal gear all over the formation ground and then ordered them to pack everything up again. The lead RI then announced that "you are going to admit to it, or one of you is going to quit." After more physical training, the RIs called on Stephen and the other Ranger candidates by name and ordered them to dance a waltz holding their rifle out in front of them. If the rifle slipped, the Ranger candidate had to empty out his gear once again and do more physical training. The purpose of this mindless abuse was to prevent Stephen and his classmates from having a few hours of sleep before getting on the buses to Camp Merrill.

At Camp Merrill, in the foothills of the Blue Ridge Mountains in Georgia, the emphasis shifts to platoon-level leadership. Here, Ranger candidates learn how to prepare their platoon's operations orders and how to plan and execute those operations in the mountains. Each mission involved rock climbing, rappelling down cliffs, and lowering military equipment down ridgelines. After three days of normal meals and sleep, Ranger candidates begin ten days' worth of twenty-two-hour workdays with only two hours to eat, sleep, and do mandatory guard duty. And the weather doesn't help. In the winter, the temperatures are normally just above or below freezing, so there is snow and freezing rain. Ranger candidates only get winter clothing—such as boot liners and GORE-TEX gear when extremely severe weather passes through the region—during all movements and operations and are only authorized their fatigues, boots, and gloves.

For Stephen, an Eagle Scout who had graduated sixteenth in his class militarily at West Point, learning the platoon operational order was not a major challenge but required intense concentration nonetheless. Each operation was planned by only three or four people at the Central Patrol Base. The platoon planners called in each squad leader one at a time to brief them verbally so they could learn their line of attack. There were no written instructions, and there was no GPS. There was only a map and whatever brief notes the planners provided. Stephen and the other Ranger candidates each had their turn as platoon planner and squad leader. RIs carefully monitored and graded every candidate's performance in each role they were called upon to play.

As an Eagle Scout, Stephen already had a lot of rock-climbing experience and so looked forward to the mountain operations. But the Merrill operations were tougher than anything he had ever experienced. Rappelling down one hundred–foot cliffs and marching long into the night without winter gear through heavy snow so no one freezes to death is not for the fainthearted. Add to that the sleep deprivation, the hunger, and the exhaustion and you begin to understand why many members of Stephen's platoon just could not carry on. Of its thirty-eight members, only seven made it through to the final stage of Ranger School at Camp Rudder in Florida. The rest were either recycled, dropped out, or eliminated from the program.

For Stephen, building supportive relationships with his fellow Ranger candidates—so important to success in Ranger School—was difficult because of the high attrition rates in his units: of the twenty-five people in the platoon he started with at Camp Darby, twelve were gone by the end of the first week. This was the normal level of attrition. But of the thirty-six he began Camp Merrill with, only seven got the go-ahead to Camp Rudder. This was unusual. At times, Stephen found the process "surreal." Suddenly, the person he had been standing next to at formation for a week was not there anymore.

The reassignments too made it harder to build up a group of supportive allies. At Darby, Stephen was assigned to a platoon in Alpha Company; at Merrill, he was moved to a platoon in Bravo Company. Everyone he had built a relationship with at Darby was no longer available as a Ranger partner.

At Merrill, he now found himself in a new squad in a new platoon in a new company. In it were several people who had previously been recycled and had a chip on their shoulder. As a result, they tried to take the leadership role on every mission, sometimes to the detriment of the squad. Luckily, Brad Hoelscher, a member of Stephen's squad, was a West Point classmate. They had known each other at the USMA and decided to work together to build alliances with others in what was now a nine-man squad. This not only helped Stephen and Brad earn good marks from their RIs but also ensured that their peer evaluations were not negatively impacted by personal indifference or hostility.

At Rudder, Stephen was again reassigned and had to build a whole new set of allies. Fortunately, he found a new Ranger buddy in Levi Diebler, a former enlisted man from the 173rd Airborne Brigade who later in his career went on to advance through the Officer Candidate School. As Stephen put it, they "stuck together like glue and figured out everything together." Levi had been recycled earlier, so he knew the other recycled Ranger candidates. This helped Stephen and Levi build the relationships necessary to get through the final Florida stage of Ranger School.

Stephen parachuted into Camp Rudder from an Air Force C17. Although it was his seventh parachute jump, he recalled that he found it "the most terrifying jump" he had ever made because he was so worried about injuring himself on landing and being recycled or kicked out of Ranger School. The professional pressure to complete Ranger School had risen to an extreme level.

Stephen was right to be worried. Unlike his previous jumps when there was thick vegetation to cushion his landing, he was now facing a hard landing. The night before it had rained heavily, and as a result the sand was now wet and hard. Stephen had a hard landing on what felt like concrete. He felt a sharp pain in his left ankle and thought he had broken it. He hadn't, but he had injured it. He detached his parachute and lay in pain on the wet sand for a good thirty seconds. A Ranger NCO came over to check on him and ask, "You okay, Ranger?" "Roger, Sergeant," Stephen replied. He rolled over, stood up, and started unpacking so the RI would go away. Then he began searching for his map and compass so he could figure out where on the vast landing zone the rest of his platoon was. For Stephen, this meant defying the pain, putting one foot in front of the other, and never quitting. Later, this was just what he did on a ten-kilometer march when blisters on his two pinkie toes burst and skin began peeling off. As he later recalled, "Pain only informs you that you are still functioning at some capacity—your character and training determines the outcome."

The start at Camp Rudder was similar to Camp Merrill. The difference was that at Camp Rudder Stephen and his fellow Ranger candidates were expected to be more advanced at this point. So, when the RIs gave Stephen and his colleagues their equipment, they expected the Ranger candidates to know how to use all of it and to secure it themselves. This was not just about securing personal or sensitive equipment but also involved mastering operational and informational security essential to mission success against any adversary.

Next came three days of mission planning and practice during which everyone ate and slept normally. This also included learning how to paddle a zodiac rubber boat and how to drag it into and out of the water. Then Stephen and his class left Camp Rudder for ten days of nonstop missions. Two took place in the Florida swamps, one out to an island off the coast of the Florida panhandle for a company-level classic World War II–style commando mission. In Stephen's platoon, the mission was to hit a radio tower on the island, while a second platoon hit a barracks and a third had to destroy a weapons cache. This was Stephen's last mission in Ranger School. The RIs canceled a final parachute jump because of bad weather.

In March 2008, Stephen graduated from Ranger School and earned his coveted Ranger Tab. It was a deeply fulfilling day for him, one of the greatest achievements of his life to date. His remarkable inner resilience and stubborn determination enabled him to successfully meet every challenge that the Army's toughest training school could throw at him. Even the RIs' unprofessional hazing reinforced Stephen's determination to succeed, but it came at a price. When Melissa came to the graduation ceremony, she almost didn't recognize Stephen: he looked haggard and emaciated, having lost nearly twenty pounds in sixty-one days. At a deeper level, Ranger School had tested Stephen like no other challenge ever had. It had forced him to dredge the depths of his capacity to endure hunger, sleep deprivation, exhaustion, and pain caused by injury. Through it all, Stephen proved he could make the right decisions and lead squad- and platoon-level missions successfully night or day on any terrain and in water. Ranger School had prepared him well for the tough challenges to come as a platoon leader in Afghanistan.

What impact does Ranger School have on you? Nick Eslinger said, "You learn so much about yourself at Ranger School, and you come out with confidence. I learned and became more self-aware of the point at which I start to drone: how far I can go without water, food, sleep, or rest before I am incapable of making good decisions. I also learned that there actually isn't a point where I could not do something. I can do it. I proved it to myself at Ranger School."

Stephen agreed with Nick's assessment but phrased it differently. "While Ranger School teaches tactics at the team, squad, platoon and introduces the company level, it does not enable you to become a tactician or expert in the operations at the company level and below. It does teach you to be the best leader in the world. It does teach you how to never quit."

16

FORT RUCKER

Fort Rucker is the "home of U.S. Army aviation." Since 1971, it is where all U.S. Army helicopter pilots have learned to fly and retrain on new aircraft or upgrade their flying skills. Fort Rucker lies amid the gentle rolling hills and woodlands of southeastern Alabama in a wiregrass county dotted with large peanut farms. It is a massive 64,500-acre base that houses a number of major commands including the U.S. Army's Aviation Center of Excellence, Aviation Warfighting Center, Combat Readiness Center, Warrant Officer Career Center, Aviation Technical Test Center, and Aviation Center Logistics Command, among others.

The Aviation Center of Excellence website summarizes the four phases that a trainee pilot must pass to graduate:

"Phase One
The first phase consists of two weeks of preflight instruction, providing students with knowledge of basic flight control relationships, aerodynamics, weather and start-up procedures.

Phase Two

The second phase, consisting of ten weeks and 60 flight hours in the TH-67 Creek training helicopter, is the primary phase. In this phase, students learn the basic fundamentals of flight, make their first solo flights, and learn to perform approaches and basic stage field maneuvers. Students then progress to more complex emergency procedure training, slopes and confined area operations.

Phase Three

The third phase is eight weeks of instrument training, including 30 hours in the flight simulator on the main post and 20 hours in the TH-67. The student progresses from basic instrument procedures to navigation on federal airways using FAA en route controlling agencies. Upon successful completion of this phase, the students are instrument qualified and receive a helicopter instrument rating upon graduation.

Phase Four

The fourth phase of training is the combat skills and dual-track phase. It is combat-mission oriented and trains the student pilot in the OH-58 A/C as an aeroscout helicopter pilot. The 1-212th Aviation Regiment teaches both tracks that include extensive night vision goggles training and tactical night operations.[1"]

17

TONY FUSCELLARO

On May 28, 2005, Tony graduated from West Point with a BSc in art, philosophy, and literature. After a little over a month of home leave, he arrived at flight school in early July 2005 to begin his eighteen-month course at the U.S. Army Aviation Center of Excellence at Fort Rucker, Alabama.

At the time, there was a logjam at flight school: too many young officers, too few instructors, and too few helicopters to train on. So, Tony and the other new arrivals started basic officership classes first. They also did simple tasks such as funeral detail for army veterans as well as other tasks around the base. Tony and the other newcomers eventually worked their way through the backlog and began flight training.

That training began with the most basic rotary-wing flight training in TH-67s, small unarmed helicopters about the size of a local television station traffic helicopter. It was an ideal aircraft to learn to fly on as well as to learn to read an instrument panel. Above all, it was simple. In the words of another Kiowa pilot, Amber Smith, the TH-67 was a "single-engine, two-blade helicopter that has the two pilots up in the cockpit and one seat for a passenger in the back. . . . It has minimal analog flight instruments, including an altimeter, airspeed indicator, vertical scale indicator, and temperature gauges. . . . [I]t had no weapons, no radar or navigational equipment."[1]

Tony's instructor was a Vietnam War veteran in his sixties who joked with him about being one of the oldest instructor pilots at Fort Rucker. Tony remembers him as someone who "had the control touch of Yoda. He was an absolute wizard in the aircraft." He also gave Tony a model of how a cool, calm pilot should behave in the cockpit.

Next up was basic navigation training in an old Vietnam-era OH58 Charlie or "Alpha Chuck," as Tony and his classmates called it. One student sat in the back writing notes while another student flew the helicopter under the direction of the instructor pilot. When flying toward a difficult landing zone that required more power, the instructor pilot would drop off the extra student in a field, where the student remained for about an hour. Tony remembers some entertaining encounters with cows curious about this human who had suddenly been left in their midst.

After basic navigation, Tony had to make his aircraft selection. As he made his choice, Tony spoke to his instructors and fellow student pilots to learn more about the four aircraft he could choose from and what he hoped to do. His choices were the Kiowa (armed reconnaissance), the Apache and the Black Hawk (ground assault), and the Chinook (transport). He chose the Kiowa and, as he put it, was "more than thrilled to select it."

Tony's choice was influenced not so much by the aircraft, its capabilities, or its weapons systems but rather by its mission: "the infantryman in the air," armed reconnaissance and close air support for troops on the ground. When Tony was in flight school, the Kiowa and its mission were the most popular choices, and you had to finish high up in your class in order to be able to select it. Tony was number one in his class and so could choose whatever aircraft he wanted.

Tony's Kiowa instructors were outstanding. Every one of them was an Iraq War veteran who taught him what to expect in combat. Among his instructors, he fondly remembers CW3 Scott Cowie, who taught him aerial gunnery. "He was a great shot and had a phenomenal love of the ground-support mission."

What was remarkable was that unlike Bobby Sickler, Tony had no formal education in mechanical engineering and still graduated at the top of his class. His success was attributable not only to his formidable intellect and fierce resolve but also because of his deep love of learning. He picked up new concepts and information quickly and was able to memorize all the necessary procedures to fly a Kiowa safely. Tony never felt overwhelmed in the cockpit. Finally, his father, Anthony, helped indirectly. Anthony gave his son his love of mathematics and engineering and provided an excellent educational foundation in both disciplines.

Tony graduated late from flight school in the spring of 2007 because he had volunteered to stay to do the Army's Survival, Evasion, Resistance, and Escape (SERE) C course that was beginning at Fort Rucker. Because of scheduling conflicts, most West Point graduates from the class of 2005 did not have the opportunity to take the SERE C course. But because he would soon be deploying into a war zone, where there was always the possibility that he could be shot down and captured, Tony felt strongly that he should stay and take this demanding course. When Fort Rucker asked for volunteers, he jumped at the chance.

SERE courses are mandatory for all Army aviation pilots. Everyone takes SERE A: a classroom-based academic course. They also take SERE B, which teaches escape techniques supplemented by a four-hour course on how to resist interrogation by enemy forces. But not everyone takes the final course, SERE C. It is the full tactical experience, a physically and psychologically stressful weeklong exercise. It began when Tony was released into the Fort Rucker training area. His task was to evade capture and use his land navigation skills to try to get to one of multiple pickup points where he would be rescued by friendly forces. At one of them, Tony was captured by the enemy. What followed was brutal. After lessons in coping with the emotional feelings triggered by capture and escape as well as effective techniques for resisting harsh interrogation, he learned when and when not to try to escape with fellow prisoners. After this came two to three days in captivity undergoing some limited exposure to torture techniques. These included being sealed in coffin-like trunks as well as boxes so small they could fit under a normal kitchen table. The psychological stress was immense. Tony got through it by repeating to himself "This too shall pass; I will be able to move on." Many years later, looking back on the experience, Tony said that SERE C had taught him that "life could always be worse. Hey, at least I'm not in a box anymore." Summing it up, Tony described SERE C as "very developmental" but added "I wouldn't volunteer to go back."

18

BOBBY SICKLER

On May 28, 2005, Bobby Sickler graduated from West Point with a degree in mechanical engineering. He took the sixty-day leave option and returned home with his family to their farm in rural West Virginia. There he spent the first part of the summer working on the farm with his father, Bob Sickler.

Although he had two months of leave, Bobby was determined to begin the next phase of his military training as soon as possible and was prepared to give up some of his allotted leave to begin flight school early. The start date for those choosing the aviation branch was determined by class rank at West Point. Bobby was not at the top of the list for the aviation branch. Luckily, those ahead of him wanted to take the full sixty days, so this opened up space in the next class, which began at Fort Rucker, Alabama, in mid-July.

Bobby arrived at Fort Rucker just after Army aviation introduced a new flight training program, Flight School XXI, and he was part of the first class to go through it. Flight School XXI's goal was to educate and train mission-capable Army helicopter pilots who were "more tactically and technically proficient" in a shorter period of time.[1] Perhaps the main reason the Army introduced Flight School XXI was that its pilots were being trained on helicopters that were not in service with the field army. As a result, Army pilots were forced to take an aircraft qualification course before they could train on the aircraft they would fly in combat: the Kiowa, the Black Hawk, the Apache, or the Chinook. Clearly, this did not make sense. Instead, Flight School XXI enabled students to learn to fly on the TH-67, a civilian version of the Kiowa. The students then made an early transition to the aircraft they would fly in combat for the bulk of their flight training.

After two weeks of classroom instruction on flight medicine, how aircraft systems worked, and what current U.S. Army doctrine was, Bobby began learning to fly on the TH-67, a simple training helicopter with the most basic analog instruments (altimeter, air speed indicator, etc.). His first instructor was exceptional: Mel Ayano, a Vietnam War combat veteran who had returned to Fort Rucker to teach after multiple tours in Southeast Asia. A quiet man in his sixties with thousands of hours in the air and hundreds of successful students, no situation surprised Ayano. For fun on weekends, Ayano raced aircraft. Bobby thought he was "a really cool guy."

Next, Bobby moved to instrument training, which meant that he spent half of each day in the classroom learning the instruments as well as the rules and procedures governing how to use them. He spent the rest of each day in a TH-67 learning to navigate his helicopter by instruments only in the clouds. Then came basic navigation in an OH-58C, the old Vietnam War–era observation helicopter. In this aircraft, Bobby's instructor taught him how to navigate using only a compass and a map as well as how to take off from and land on rough terrain.

At this point, Bobby had to choose his primary aircraft. He had originally wanted to fly Apaches, but when the Army's aviation command gave the students in Bobby's class their options, there were only two: Black Hawks and Kiowas. There was no opportunity to choose the Apache. So, Bobby chose to fly Kiowas because he wanted to be "danger close" to combat protecting and supporting the infantry. "The mission of the scout helicopter, scout attack, and reconnaissance appealed to me personally," Bobby told me. "It fit my personality and what I wanted to do."

To be able to choose the Kiowa, you had to have performed exceptionally well in flight school. Bobby did; he was ranked second in his
class. He had struggled at West Point because West Virginia's underresourced schools had not prepared him for the USMA's Ivy League–
standard academic courses. But his raw intellectual firepower and steely
resolve enabled him to succeed and graduate with a good GPA.

Now Bobby's rigorous West Point education in mechanical engineering gave him a real advantage in his academic courses at Fort
Rucker. For example, he had covered most of the flight school engineering course material at the USMA. Bobby had a second advantage.
Thanks to his boyhood flights with his father, Bobby was already familiar with instruments, basic navigation, and weather patterns. His father
gave him a strong intuitive feel for flying. Finally, growing up on the
family farm where he had learned to operate different types of machinery and ride a motorbike also helped. This familiarity with learning and
operating machinery helped Bobby learn to fly a helicopter faster than
most of his Fort Rucker classmates. As Bobby later recalled, "To me,
flying was just operating another machine." By the end of flight school,
Bobby was an honor graduate.

When Bobby learned to fly the Kiowa and took his advanced combat training in it, he was blessed with outstanding instructors. They were
a mix of civilians who had flown Kiowas for many years and Army warrant officers who had just returned from fighting in Iraq and Afghanistan. The civilian was Dudley Carver, "a legend in the Kiowa community,
the crazy old warrant officer," as Bobby later described him. The young
warrant officers were Dan "Combat Dan" Beurachua, in his late twenties
who had gone straight from high school to flight school, and Joe Beebee,
in his early thirties, who had been an enlisted man before going into the
Army's warrant officer program and had just returned from Iraq. They
were all outstanding teachers who taught Bobby what to expect going
into combat in the Kiowa.

After flight training, Bobby completed SERE A and B as well as
the Aviation Captain's Career Course and graduated second in his class
from Fort Rucker in October 2006.

PART 4

INTO BATTLE

19

NICK ESLINGER

Iraq, 2008–2009

Nick Eslinger's next challenge was to become an effective platoon leader. In mid-May 2008, Nick reported to his first duty station, the 2nd Battalion, 327th Infantry Regiment, First Brigade, 101st Airborne Division, at Fort Campbell, Kentucky. When Nick arrived at Fort Campbell, he found that his entire brigade had deployed to Iraq. He was now on rear detachment awaiting deployment. The 101st Airborne Division policy was that all new officers and soldiers had to go through a month of individual readiness training in preparation for deployment: medical training, marksmanship qualification, vaccinations, and procedural tasks. Every day felt like a week: Nick not only wanted to take command of his platoon and prove himself as an infantry officer in combat but also wanted to get to Iraq before conditions on the ground changed and the war might end.

On June 27, 2008, Nick boarded a contracted flight bound for Kuwait. There, he spent another week conducting mandatory training before going into Iraq. On July 4 he flew into Tikrit, Iraq, where the First Brigade headquarters was located. As Nick recalls, "I was that cheery lieutenant running around brigade headquarters asking when the next convoy was leaving for Samarra. Thankfully a resupply convoy had arrived from Samarra, and I was able to join them for their return. The convoy consisted of two MRAPs [Mine-Resistant, Ambush-Protected vehicles], six soldiers and me."

Thus, Nick's first journey "outside the wire" was in a vulnerable convoy with no armor and very little security. The travel time from Tikrit to Samarra was ninety minutes. On July 5, 2008, Nick reported to his battalion commander, Lt. Col. Joseph McGee (West Point class of 1990).

Lieutenant Colonel McGee was a charismatic, courageous officer who exuded leadership. His policy was that every new officer who joined his battalion had to spend a few days shadowing him. During these two days, McGee asked questions as he informally sized up each officer. As Nick and another new officer traveled with the commander in his MRAP to and from different checkpoints in Samarra, he kept asking questions. On one mission on a Friday as they drove through a village, McGee asked, "Why do you think none of the locals are out today? Where is everybody?" Neither Nick nor the other lieutenant could answer correctly. Nick didn't know what McGee was getting at and felt stupid that he didn't know the answer. McGee gave him the answer: "Friday is a day of rest for the Iraqis, and everyone is in their homes. No one works on Fridays." In the end, McGee did not hold Nick's lack of knowledge against him.

Lieutenant Colonel McGee dominated any room he walked into, not because of his rank but because he was such a revered commander. Earlier he had spent several years in the Army's elite Ranger Regiment, and this commanded respect throughout the battalion. A dynamic and inspirational officer, McGee knew how to motivate his soldiers to fight throughout their long, dangerous fifteen-month deployment. He never isolated himself in his headquarters. Rather, he was always out with his soldiers on missions or meeting local tribal leaders or the mayor of Samarra. McGee had all the qualities of a transformational leader:

he was superbly competent, he always led his soldiers by example, and he engaged effectively with Iraqi partners as well as local Iraqi tribal leaders, all of whom respected him. Not surprisingly, he made real strides toward achieving General David Petraeus's desired outcomes throughout the battalion's operational area in Samarra: improved security, better law enforcement, fixing the failing electrical grid, and improving agriculture.

Lieutenant Colonel McGee knew his battalion intimately. He knew the personalities in each platoon and what kind of leader each platoon needed, that is, what type of personality would fit best with each platoon. Two days later, McGee called Nick into his office at 1 a.m. and said, "Nick, today at 6 a.m. there is going to be a convoy from Cougar Company. They will arrive here at our FOB, and you are going to go down to Samarra with them. You're going to be Cougar One Six." Nick replied, "Awesome, sir, thank you. I'm ready to go." Nick traveled down to Patrol Base Olson, reported to the acting company commander, and met his platoon. The company CO, Capt. Josh Kurtzman, was on leave.

Patrol Base Olson was a small base, home to about 250 U.S. soldiers. They comprised Nick's company, Cougar Company, a small group of special forces troops and maintenance and logistics personnel. The camp itself was about the size of three football fields put side by side. The largest structure on the base was a two-story wooden building with a basement totaling between six thousand and eight thousand square feet. In addition, there were also a couple of one-story wooden structures for conference rooms as well as a gym. The soldiers lived in containerized housing units. Patrol Base Olson also contained a parking lot for the vehicles, a fuel point, and a helicopter landing zone. The perimeter consisted of Hesco barriers and a heavily armed force protection tower on each of the four corners.

Cougar Company's area of operations was the small city of Samarra, 128 kilometers north of Baghdad near the tip of the notorious Sunni "Triangle of Death." Samarra, an ancient holy city of Islam on the east bank of the Tigris River, has a long, rich history as a center for Islamic scholarship and worship and a destination point for Sunni Muslim pilgrims. The city is home to the Golden Mosque, a major Muslim religious site famous for its spiral minaret. More recently, however, Samarra had become one of the centers of the Sunni insurgency, a trouble spot for U.S. forces. Two years before Nick arrived in Samarra, a minaret of its ninth-century mosque had been blown up, triggering protests and violence throughout Iraq. The mosque's remaining minarets were destroyed early in 2008. For a young West Pointer and his platoon, Samarra was hostile territory.

In size, Samarra is only three miles (east to west) by two miles (north to south) and had a large population of a quarter of a million people. When Nick arrived in September 2008, Samarra was a grim war-torn city: the walls of most buildings in the city were pockmarked by bullet holes and blast residue. The city's main streets were asphalt punctured by numerous potholes from previous IED explosions. The side streets and alleyways were unpaved. Most of the buildings were between one story and three stories and were made of yellowish adobe clay. Samarra's only tall building was the six-story hospital. In the city's more affluent residential areas you could find stucco buildings, but they were rare.

Samarra's electrical grid was a mess. Electrical wires were exposed everywhere. When on patrol, Nick's platoon often had to slow down so the gunners could lift the wires over the turret and lower the .50-caliber machine gun so their MRAP could move safely forward. Outside the city, there was farmland and desert to the north and farmland to the east and south. The farmland was parched and poorly irrigated. Among the population of Samarra, women and military age men were stand-offish, skeptical of the U.S. military presence. In contrast, the children were warm and welcoming.

If Lieutenant Colonel McGee was an outstanding battalion commander, Capt. Joshua Kurtzman (USMA class of 2001), Nick's company commander, was equally extraordinary. At that time, Captain Kurtzman had more command experience in combat at the company level than any other officer in the Army. He had taken over command of Cougar Company during its previous deployment, continued during the train-up for this one deployment, and remained in command for this fifteen-month deployment to Iraq. He was reserved and quiet, but when he spoke everyone listened.

When Nick arrived to join Cougar Company, Capt. Kurtzman was on leave, so it was ten days before they met. During that time, Nick's soldiers told him what an amazingly selfless man their company commander was. Soldier after soldier described Kurtzman as a superhero. When Kurtzman returned from leave and Nick had the chance to get to know him, he too described him as a superhero. The reason everyone in Cougar Company respected and revered Kurtzman so much was his humility and extraordinary selflessness. He always put the well-being of the men of Cougar Company ahead of his own. He slept only a few hours per night because he was always working on behalf of soldiers or their families. In Nick's words, Kurtzman was "always on a mission, the most selfless man I ever met." In Nick's view, Lieutenant Colonel McGee and Capt. Kurtzman were the kind of exceptional officers he wanted to emulate. "They exuded leadership in the way that I envisioned the perfect leader should."

The next task for Nick was earning the trust of his soldiers. He was determined not to rest on his laurels as a West Point and Ranger School graduate, much less be a self-promoting ring knocker. Nick approached the task with confidence but knew that he faced a difficult challenge: the platoon was already deployed in Iraq, and he had had no opportunity to get to know his soldiers as he normally would during the pre-deployment train-up. Nick wanted to demonstrate that he was a caring, selfless officer who would always put their well-being ahead of his own and would never ask them to do anything he was not prepared to do himself. He wanted them to know that he would do whatever it took to look after his soldiers and accomplish the mission.

To do this, Nick knew that he needed to earn the confidence and respect of his platoon sergeant and squad leaders. They were the established leaders in the platoon who had already earned the trust of the soldiers. If he could win their confidence, the rest of the platoon would follow. Thus, on his first day in command he called his platoon sergeant, SSgt. Christopher Anderson, and squad leaders, SSgt. Miguel Hendry, SSgt. Samuel Heath, and SSgt. Jared Gass, into a meeting. After introducing himself and explaining his near-term expectations, Nick said, "I would like to address our platoon today, but I would appreciate your feedback on what you think they need to hear from me right now, given the circumstances of Sergeant First Class Chevalier's death yesterday." Sfc. Steven J. Chevalier was the platoon sergeant in the 3rd Platoon, Cougar Company, and had died on July 9, 2008, of wounds sustained when his vehicle was struck by a grenade in Samarra. Nick listened carefully as his squad leaders advised him to empathize with the soldiers who were hurting over Chevalier's death, maintain a steadfast focus on our task and purpose for each mission, and not make any major changes in the first few days. Nick accepted all of their valid recommendations. His first address to his soldiers was effective because it was infused with the recommendations of his NCOs. After that, Nick always respected the opinions of his NCOs and acted on their recommendations if they were valid. He rarely overrode them, nor did he criticize them. He just tried to be the most humble and selfless leader he could. On patrol, Nick always led from the front, always in the most advantageous and dangerous position on the ground, never riding in the air-conditioned comfort and relative safety of his platoon's armored vehicles.

Throughout the summer months, Nick constantly sought feedback from his NCOs and did everything a good platoon leader should to solidify the respect he had earned. He succeeded not only because of his humility and selfless courage on October 1, 2008, but also because West Point prepared him so well to lead troops in combat. He had learned how to successfully navigate the complex dynamics of the officer-NCO relationship and how to inspire and motivate soldiers.

On their two patrols a day, one in daylight one at night, Nick and his platoon remained vigilant, doing everything a disciplined infantry platoon should do: maintaining sectors of fire, adhering to noise and light discipline, using brevity on the radios, adhering to escalation of force procedures, and constantly scanning the immediate area.

October 1, 2008

On the morning of October 1, 2008, Nick led an uneventful five-hour patrol before returning to Patrol Base Olson. The goal of the platoon's mission was to gather intelligence about some recent RKG3 antiarmor grenade attacks on U.S. vehicles.

Once back in the relative security of Patrol Base Olson the platoon ate, worked out in the gym, and rested. Nick and his platoon sergeant planned their night patrol. They briefed the men and began to carry out their normal prepatrol procedures: soldiers prepped vehicles, their gear, and their weapons. Everybody knew their job and did it well. The platoon had been in Iraq for ten months, Nick for three months, so the platoon commander and his soldiers had now begun to know each other well.

The evening grew cold, with the temperature falling to the low forties. In the twilight, Nick led his second patrol of the day into Samarra's dangerous, densely populated Jiberia 2 neighborhood. Until now, he had experienced neither direct enemy fire nor any form of combat with the enemy.

That evening, as in the morning, the mission was to gather information about some recent RKG3 antiarmor grenade attacks on U.S. troops. At 7 p.m., Lieutenant Eslinger and the thirty-six members of his platoon climbed aboard their armored Humvees. They left Patrol Base Olson in downtown Samarra, driving along the bank of the Tigris River.

On their way, they picked up a squad of Iraqi national policemen. This was to be a partner patrol. The patrol's goal was to return to the Jiberia 2 neighborhood to follow up on leads from their morning patrol. At about 7:30 p.m., Nick's platoon reached Jiberia 2's narrow alleyways and multistory buildings. Nick dismounted with some of his men. As per normal operating procedures, his platoon sergeant stayed with the mounted element in the MRAPs to provide a screen around their dismounted movement route.

Nick had a preplanned dismounted route. Along it, he planned to knock on doors, talk to a few people, and see if he could get any actionable information. At the first house, Nick and the Iraqi national policemen drank tea with the homeowner and had a general conversation.

At about 8:30 p.m., Nick and the Iraqi policemen reached the second house, their primary target for the patrol. There was still ambient light. As a result, Nick did not put his night vision goggles on. Unbeknownst to him at the time, that decision may have saved his life and the lives of several of his soldiers. The narrow alley the platoon now occupied was only wide enough for one or two vehicles to safely maneuver in. It was lined with tall yellow-brown mud walls. To Nick's left was an L-shaped wall, and to his right was another tall mud wall. In front was an open courtyard, a building site with clumps of dirt lined by a mud wall on three sides. It was a square lot. On its back side was an eight-foot mud wall, the backyard wall of another house.

In the narrow alleyway, Nick and the dismounted element of his platoon came to a halt while SSgt. Sam Heath got the Iraqi police squad leader and walked to the courtyard door of the second targeted house. As he did so, Nick and the other dismounts close by—Specialist Courson, his radio tactical operator (RTO); Specialist Crowell, the squad's M249 Squad Automatic Weapon (SAW) machine gunner; and Private Benjamin—all pulled security. There was only sixteen feet spacing between them all. In a tight alleyway barely thirty-two feet wide, there was very little space to maneuver. To the soldiers' left and right were tall yellow-brown mud walls, the outer walls of the small courtyards in front of each house. The alleyway was so narrow they could barely get their MRAPs down it.

As Nick and his dismounted soldiers pulled security, Staff Sergeant Heath and the Iraqi police squad leader knocked on the door. To Nick's left was a tall L-shaped mud wall with a 90-degree turn. To get some cover, Nick positioned himself against the wall with his eyes focused around the corner. In front of him was an open courtyard, a dirt lot about sixty-five feet square with lumps of earth lined on three sides by six-foot-high mud walls. On the back side of the lot was the backyard wall of another house.

When Staff Sergeant Heath knocked on the door, a man answered and spoke to him through Alex, Nick's interpreter. At this time Nick had raised his hand up to his left shoulder to press the radio button so he could speak quietly over the radio to Heath. Nick wanted to remain as silent as possible as he spoke. He whispered into his microphone "1-3 this is 1-6. Are we good to enter?" As he looked up, he saw a shadowy hand throwing something over the wall. At that moment, there were five American and two Iraqi lives directly at risk. Right next to Nick was Specialist Courson. Across the narrow alley were Staff Sergeant Heath; Private Benjamin, who was helping Heath; Nick's Iraqi interpreter, Alex; and the Iraqi national police squad leader. To Nick's immediate right was Specialist Crowell, a rifleman helping him pull security. Nick and Crowell were the two responsible for pulling security on the open lot. Not too far away was the platoon medic, Private First Class "Doc" McMillan. The rest of the squad and the remainder of the platoon spread down the narrow alleyway.

Instantly, Nick recognized the threat as a hand grenade. He didn't stop to think what to do. Instead, driven by a finely honed selfless instinct to protect his soldiers, he got up off his knee and threw himself six to eight feet to his right trying to catch or block the grenade. As he landed on his right side, the grenade landed and rolled into his body armor near the right side of his ribs. As he recalled,

I felt the plunk when it rolled into me; it stopped immediately, as if wedged between my vest and the ground. I rolled my body to the right to cover the grenade as I tried to bring my hands in to grab it. I got my right hand on it and then quickly rolled and flung it toward the wall it came from. I just grabbed it and threw it as hard as I could, which wasn't very hard because of my body position. I tried to yell "grenade" as I threw it, but before I could get to the end of the second syllable, the concussion of the explosion took my breath away. I didn't even have time to put my head down.

The grenade exploded in midair. Luckily, it exploded high enough so the shrapnel from it flew above the heads of Nick and his platoon.

Nick knew he wasn't hit, so his first thought was for the safety of his soldiers. He asked Specialist Courson whether he was hit. He replied, "No sir." Nick said, "Get Doc," the platoon medic. It was now about five seconds after the concussion of the explosion. In the alley, there was no movement, no sound. There was a moment of crystal-clear silence, something Nick had never experienced before. In his mind he thought, "Whoa, that just happened. Oh my goodness."

Doc McMillan came running over and started padding Nick. "There's no way you're not bleeding right now, sir," he said. "No, I'm fine, Doc," Nick replied. "Look at me. Go to every person on the ground. Check them and report back to me." At the same time, Nick told his RTO, Specialist Courson, to radio the platoon sergeant to bring the vehicles up for security and evacuation of any casualties. Ever vigilant, Staff Sergeant Heath had not seen the grenade being thrown but immediately turned around at the noise of Nick landing on the ground. Later he told Nick, "I thought you were wrestling with somebody. I didn't know what the hell you were doing, sir; you just looked like you were wrestling on the ground with somebody."

On Nick's orders, Staff Sergeant Heath took a team south down the narrow alley trying to find a way to go east and then north to try to find the enemy insurgent who had thrown the grenade. At the same time, Doc McMillan returned to tell Nick that miraculously not one of the twelve dismounted soldiers, the Iraqi interpreter, or the Iraqi national police squad leader were hit. He just couldn't believe it. After reporting to Nick, McMillan ran off to catch up with Staff Sergeant Heath.

The next thing Nick heard was the noise of three MRAPs screaming down the alleyway even though they could barely fit. They screeched to a halt one behind the other. Nick shouted to his platoon sergeant "You need to move; this could be a trap. This is an ambush point. We can't bring our vehicles through here. Keep only one here. Have the gunner scan that sector. I'll have the SAW gunner scan this sector. Move the other vehicles to an isolation position."

Nick's RTO then called up the company command post at Patrol Base Olson: "Cougar Main, this is 1-6 Romeo. Do we have air assets in the area?" His goal was to get some eyes in the air to check if they saw any enemy activity that he and his soldiers could then locate and engage. The company command post replied that the only air assets available that night was a pair of Air Force F-16 fighter jets flying back to their base for the night. So, Nick requested and got a show of force: a high-speed, low-level "scream past." It was a spectacular show, but unfortunately the Air Force pilots did not see anything, and there were no helicopters available to provide airborne surveillance.

After fifteen minutes, Nick and Staff Sergeant Heath realized that they weren't going to find the insurgent who threw the grenade that night. But if they stayed where they were and questioned the owners of the surrounding houses over whose back wall the insurgent had thrown the grenade, they might develop some leads.

Nick set up a security perimeter, and he, his Iraqi interpreter, Alex, and Staff Sergeant Heath went into the house. Inside, there was a middle-aged woman but no men, and that immediately aroused Nick's suspicions. Through his interpreter he began asking the woman general questions: "Did you see anything? Where is your husband? Do you have any sons? Why are there no men in the house?" As she began to answer, Nick got a radio message from one of his squad leaders. "Hey, sir, you probably want to come in the backyard and take a look at this."

Nick went out and found a crude drawing of an ambush on the side wall of the house. He and his accompanying soldiers shone their flashlights on it. They could see what looked like ancient Native American hieroglyphics. There was a crudely drawn picture of a narrow alleyway. In the middle was a big vehicle that looked like an MRAP with American soldiers walking in front of it. In the next frame was an explosion drawn to look like fire. Nick and his soldiers were stunned. The drawing depicted exactly what had just happened to them a few minutes earlier.

To Nick this was evidence that the ambush had been planned here in this backyard by someone who lived in this house. It was a lead that had to be followed up immediately. He went back into the house and intensified his questioning of the middle-aged Iraqi woman. After about five minutes, she admitted that it was her son who had made the drawing and carried out the ambush. Nick asked, "Where is your son?" She replied, "I don't know." Nick got the son's full name and as much additional information as he could. Most important, Nick learned that her son was a member of the Sons of Iraq. So, when Nick learned this, he thought, "Okay, we'll find him tomorrow if he reports for work." Nick ended the patrol for the night. The platoon remounted their vehicles and returned to Patrol Base Olson, where Nick did his mission debrief. The next morning his platoon successfully detained the individual who threw the grenade the night before.

In Samarra, Lieutenant Colonel McGee's battalion had three different types of Iraqi security forces to help them in securing the city: the police, the army, and the Sons of Iraq, a new paramilitary organization just beginning in the city. The organization had begun in 2007 in Anbar Province as part of General Petraeus's new counterinsurgency strategy. There, as Michal Harari suggested, "Coalition forces succeeded in co-opting tribal leaders, alienated by Al-Qaeda's extremist ideology and brutal tactics, and 'turned formerly passive supporters as well as some former insurgents into active supporters of the counterinsurgency effort.' They convinced tribal leaders to recruit young men, mostly Sunni-Arabs, and create volunteer security forces around the country."[1] So, instead of killing American and other coalition forces, these men were now paid and trained by U.S. forces to help suppress the violence. On the ground in Samarra, the Sons of Iraq ran routine checkpoints.

Back at Patrol Base Olson, Nick was still pumped full of adrenalin after his first direct experience of combat with the enemy. With characteristic humility and professionalism, he kept replaying in his mind everything he did after the insurgent threw the grenade. What had he done right? What mistakes, if any, had he made? Nick wanted to capture any lessons learned and kept asking his platoon sergeant "What could I have done better?" But all Sergeant Anderson kept saying was "Are you okay?" He was trying to assess what kind of emotional response Nick was having to the insurgent attack, the results of which could have been catastrophic. At that point, the platoon sergeant was more concerned with Nick's personal welfare than giving him tactical feedback. Nick now wanted to refocus on tomorrow and to review routines and procedures. Before he could do so, he had to meet his battalion commander, Lieutenant Colonel McGee, who had asked to see him.

Lieutenant Colonel McGee asked Nick to tell him what had happened, which he did. McGee then said, "Can you tell me a little bit about why you did what you did?" Nick replied that it was a reflex, an instinct to save his soldiers' lives, and that he had no time to deliberate options before the grenade. McGee asked him some additional questions, including why he had moved toward the grenade and not away from it. As their conversation progressed, Nick—with characteristic humility—began to think he had done something wrong. "Should I have just yelled grenade and dove the other way?" he asked McGee. The battalion commander was noncommittal. Nick left the meeting feeling down on himself and worried that he had disappointed his battalion commander or, worse, had put his soldiers at unnecessary risk due to his actions that night.

Later Captain Kurtzman told him not to worry; he had done nothing wrong. Lieutenant Colonel McGee was simply trying to get a clear picture of what had happened because he had to determine what level of valorous award was warranted, if any. Nick didn't spend a moment thinking about awards, but he was glad Captain Kurtzman had told him because it stopped him from second-guessing himself and enabled him to get back to being a normal platoon leader. As to a medal for valor, Nick's mindset was "Leaders will decide. Either way, there were a hundred more important things going on that required my attention." But McGee's questions really made Nick think: "Why did I do that?" Years later, Nick reflected on his actions:

> Truly, it was a reflex. And when I think about that reflex to move towards it and not away from it, I really think it comes from developing the state of courage through West Point, through Ranger Schools, just the experiences I have had ever since I left high school leading up to that night in Iraq. It just developed a state inside of me

that influenced my behavioral reaction to a grenade being thrown at me. That state of courage is multidimensional, and it has a lot to do with being selfless and being a servant: understanding and believing that a leader is a servant and your soldiers' lives are more important than your own. Your needs are subordinate to theirs, and you just do anything for them. I call it professional love.

To that, I would add that this professional love was built not only by West Point and Ranger School but also by the remarkable selfless values taught and practiced by Nick's parents, Donna and Bruce. And they were reinforced by the values championed and practiced by Captain Kurtzman and Lieutenant Colonel McGee.

The last seven weeks of Nick's deployment were calm. Neither he nor any of his platoon had contact with the enemy. Nick attributed it to the method of T-walling around Samarra.[2] A T-wall is a twelve-foot-high by four-foot-wide cement wall. You can line T-walls up so that they interlock with each other. Nick's battalion's mission was to put T-walls around the city of Samarra. Every night a different platoon would go out with about thirty flatbed trucks. Each carried ten T-walls. An Iraqi crew of crane operators met the U.S. soldiers at an agreed location in the city. The soldiers pulled security while the crane operators put the T-walls in place.

By putting in the T-walls, the battalion corralled insurgents. The T-walls steadily closed off insurgent access to and from Samarra. This had three good results. The first was to stop insurgents from coming into the city. The second was that the T-walls enabled U.S. forces to locate insurgents inside Samarra more quickly. The third was that the T-walls encouraged threatened insurgents to leave the city. Taken together, these three results contributed to great security in the city and to a reduction in the number of insurgent attacks on U.S. forces.

Nick believed that the other thing that contributed to greater calm in Samarra was the Sons of Iraq program. That program gave young military-age men an alternative to joining the insurgency. Instead of planting IEDs for fifty dollars a time, they could join the Sons of Iraq, get paid regularly, and feed their families. In and around Samarra, the Sons of Iraq ran tactical checkpoints and helped relieve some of the security burden on U.S. troops. There were too many streets and foot routes for Nick's battalion to manage on their own. With the Sons of Iraq taking over routine duties, the battalion could focus their efforts on the neighborhoods.

20

TONY FUSCELLARO

Afghanistan, 2009–2010

In the summer of 2007, Tony Fuscellaro joined his first operational unit, the First Squadron, Seventeenth Cavalry Regiment, 82nd Combat Aviation Brigade, 82nd Airborne Division. They were returning from their deployment in Mosul in 2007. Lt. Col. Mike Piat was in command but was nearing the end of his tour with the squadron. His successor had already been chosen: Lt. Col. Mike Morgan, a twenty-two-year Army veteran who had already deployed six times to Iraq and Afghanistan. He took command of the squadron in late October 2007.

A native of Virginia Beach, Lieutenant Colonel Morgan had been a star soccer player for Kempsville High School. After qualifying as an Army helicopter pilot and serving with distinction in the regular Army, he had been transferred to the Army's elite 160th Special Operations Aviation Regiment, where he had spent many years flying "Little Birds." The "Little Bird" AH6 is a small highly maneuverable helicopter with a cabin shaped like an egg. Nicknamed the "killer egg," its main mission is to insert special forces commandos onto rooftops or other narrow spaces and give them close air support with its 2.75-inch rockets, hell-fire missiles, and heavy machine guns. The attack lift variant carries up to four special forces commando operators on benches fitted to the doors on either side of the aircraft. The attack variant, which Lieutenant Colonel Morgan flew, was equipped with a 7.62-caliber minigun and a 2.75-inch rocket pod on each side of the aircraft.

Lieutenant Colonel Morgan's background was similar to Tony's in two important ways. The first was that they both had exceptional parents who taught them the values of humility, hard work, and selfless service to the nation and to others. The second similarity was that they came from blue-collar families.

A highly decorated officer, Lieutenant Colonel Morgan was also a dynamic, inspirational leader: a humble, self-deprecating man with great presence, never shy about learning from his subordinates. His interpersonal skills were remarkable. A humble man, he respected every soldier under his command and could connect with the soldiers on a very personal level. The sincerity and compassion he showed to every soldier in the unit endeared him to his troops.

At the same time, Lieutenant Colonel Morgan held everyone to the highest standards of professional excellence, himself most of all. He inspired everyone else in the squadron to execute their military duties to the highest level possible. Morgan was someone who carefully balanced the serious business of being a professional soldier and the unit's CO, with love of the mission and the ground forces. In his mind, the only reason Army aviation existed was to support and save the lives of the troops on the ground. As Tony later recalled, "He brought that view with him from day one." Morgan, knowing that he would deploy the unit to Afghanistan in the summer of 2009, almost immediately began to change the squadron's culture from that of the regular Army to something closer to the elite 160th Special Operations Aviation Regiment.

To that end, Lieutenant Colonel Morgan began with the commissioned officer pilots. He made it clear that he wanted every pilot in the squadron to be extremely proficient in the aircraft, to strive to be the best possible professional they could be. Prior to Morgan's arrival, the accepted culture and practice had been that the noncommissioned warrant officers far exceeded the commissioned officers in flight proficiency and flew the aircraft. Morgan disagreed and ensured that every pilot was held to the same standard. He also ordered that commissioned officers learn their aircraft thoroughly: how to refuel and rearm the aircraft, including loading rockets into their pods and the 7.62-round bullet belts into the machine-gun chutes. This was not to be left solely to the ground crews. Every second counts in combat when soldiers' lives are at risk. Morgan wanted what he called "NASCAR-level efficiency in every fuel or ammo stop."

Next, Lieutenant Colonel Morgan wanted pinpoint accuracy from his pilots. He was an exceptional pilot with vast combat experience who wanted to raise his squadron's gunnery standards to the same level he had experienced in the elite 160th Special Operations Aviation Regiment. He did not believe in proximate accuracy. Driven by his passion to save U.S. soldiers' lives, Morgan drew on his special operations experience and brought their exacting flexible techniques of warfighting to regular Army aviation. He even recruited borrowed NCOs from the elite 160th to recalibrate his squadron's weapons to match the techniques used by AH-6 armament specialists.

The commander taught his men that the rockets fired from their pods on either side of the Kiowa crossed between three hundred and four hundred meters after leaving the aircraft. So, the ideal distance to engage the enemy was five hundred to seven hundred meters. It's always a close fight. His message to the squadron was that the Taliban were first-class guerrilla fighters and that to defeat them the squadron would have to fight them in a more innovative way than existing Army doctrine prescribed.

The key to success was pushing the unit to achieve pinpoint accuracy when the soldiers pulled the trigger. When the majority of the targets are surrounded by sand, the blast radius is incredibly small. Hitting the target was the only way to guarantee success. To accomplish that, the squadron needed endless practice with an unlimited supply of ammunition. Lieutenant Colonel Morgan demanded and got the ammunition he wanted. This too was a big change in the unit's culture and organizational practice. Prior to Morgan's arrival, test-firing was done simply to ensure that weapons were working properly, which meant one rocket and one burst from the machine gun. During the 2009 deployment, Morgan ensured that every crew fired at least four to seven rockets and one hundred rounds to identify the impact points of the weapon systems and to "calibrate your eye to the aircraft."

As Tony recalled in his speech at West Point accepting the Nininger Medal on September 16, 2013, Lieutenant Colonel Morgan's

> approach to aerial marksmanship was exacting and demanding. He expected every round to be a direct hit. For example, if we were shooting at a vehicle, we would call out the tire or window for which we were aiming. If the vehicle was hit but the aim point was not hit, then your shot was a failure. As we became more skilled, we would toss a colored smoke canister out the door and hit the six-inch piece of tin before the smoke dissipated. The standard was well above what was required in manuals. In fact, in most cases the training standard exceeded what was required in actual firefights.

Essential to this task was mastering what Tony described as the "quirks" of the .50-caliber machine gun. Each one performed differently under combat conditions. Some tended to oscillate in a circle, while others started low and then dragged the rounds up and right as you flew. Each gun was slightly different depending on how they were mounted on the airframe. Lieutenant Colonel Morgan demanded that every pilot in the unit know their gun so that when they opened fire they knew where their first round would go.

This intensive practice was vital to the success of the upcoming deployment, because unlike Apaches and Black Hawks, the Kiowa had no computer systems, no laser-guided missiles, and no advanced technologies to tell the pilots where to fire. Tony described the Kiowa as "a 1980s Porsche." Deciding where to shoot was largely manual and required "Kentucky windage," as Lieutenant Colonel Morgan used to teach. Kiowa pilots lined up their weapons and their targets "by putting a grease-pencil mark on the windshield, then lin[ing] up their target with the mark."[1] Morgan compared it to nineteenth-century shoot-outs in the Old West: "It's like gun fighting in the street with a six shooter. . . . It's very old school, but it works every time when you need it most."

This was a demanding but exciting culture within which a young officer such as Tony could grow and develop under the leadership of an inspirational leader such as Lieutenant Colonel Morgan. The squadron was not due to deploy until the summer of 2009, so Tony now had eighteen months to integrate into the squadron before its next deployment in harm's way. His biggest challenge was how to build bonds of trust with a group of pilots who had fought together in Iraq and built those unique bonds that develop between combat veterans. There was a gap between those who had deployed and those who had not, and it took time to break through that wall.

Tony's approach was "competence first, personality second." With single-minded determination, Tony set out to show his comrades that he was not only a competent pilot but also an officer who could bring real value to the squadron. As he proved his competence as a pilot and platoon leader, he found that the squadron was open to accepting him.

By November 2008 Tony completed his platoon leader time. Lieutenant Colonel Morgan had observed Tony's talents as well as his fierce determination to be the best pilot he could be and had chosen him to be his personnel officer and adjutant. Morgan also saw that Tony was a good writer and had a real empathy for people and their goals and relationships. The young captain lieutenant and his CO soon developed a close professional relationship.

At that same time, Tony had a concern. Switching from being a platoon leader, just as he was building trust with combat veterans, to being a staff officer was risky. Instead of going into combat with his squadron as a line pilot poised to fly every day, there was the possibility that he would miss the deployment now that he was a staff officer with other duties and requirements. It was a big transition, and he feared missing flight opportunities.

Tony need not have worried. In fact, becoming Lieutenant Colonel Morgan's adjutant would prove to be the most important part of his development as a company-grade field officer. Working alongside Morgan was a wonderful learning experience for a young officer like him. From Morgan, Tony learned the importance of a relentless commitment to professional excellence and continuous skill development and also learned to hold yourself to higher standards than you hold your subordinates. He learned how to be known by your skills; to never brag; to show humility, sincerity, and compassion in dealing with every soldier; and to treat everyone from privates to generals with the same respect.

At the same time, Lieutenant Colonel Morgan saw Tony's potential and had begun to groom him to be his wingman when the squadron next deployed. Morgan saw that they thought the same way and had similar approaches to military leadership. He also saw that Tony had great potential as a combat pilot and wanted to teach him the advanced combat flying skills he would need in Iraq or Afghanistan.

Lieutenant Colonel Morgan's decision to groom Tony as his wingman for the squadron's next deployment was part of his wider leadership philosophy about combat flying. Morgan believed that pilots in a Kiowa squadron should always fly with the same people to build trust and greater combat effectiveness. The opposing view in the Army aviation community was that teams should be mixed up periodically to ensure that they did not become stagnant or complacent. The squadron's combat experience in Afghanistan in 2009 would prove Morgan right.

Although Tony was serving as the Lieutenant Colonel Morgan's adjutant, he was also his wingman and learning the advanced combat skills he needed to be effective and to survive. Morgan described armed reconnaissance and close air support in Afghanistan as "a knife fight" in which you are head-to-head with your enemy. Kiowas are always in danger, he told Tony, always close to combat or in combat, always at risk. It was therefore vital for Tony to know his aircraft and to fight effectively with it.

The Kiowa was small, fast, and remarkably agile. It presented the Taliban insurgents on the ground with a difficult target to hit. The aircraft itself had a tiny, cramped cockpit with hard, uncomfortable seats and very limited armor protection: only a narrow piece of steel on each side of the cockpit covering most of the side of each of the two pilots except their heads. The Kiowa had no doors, no heating, and no air-conditioning. For the pilots, the temperature inside the cockpit was always too hot or too cold. Despite its small size, the Kiowa was well armed with one .50-caliber machine gun and two pods carrying seven rockets each. Alternate configurations included two seven-shot rocket pods without the machine gun or a Hellfire missile rack with two missiles in place of a rocket pod.

In addition to speed, small size, and maneuverability, the Kiowa had other features that were to prove invaluable to Tony in Afghanistan: a very large windscreen, a small console, and a transparent plexiglass chin bubble at his feet beneath the pedals. Together, they gave Tony enormous situational awareness. As he described it, "I could see every single soldier and see everything from their perspective I could see directly over the shoulder of the ground forces." Later in Afghanistan when soldiers under enemy fire from a hut or *qalat* (fortress) would radio for help, Tony always knew and could quickly decipher where they were and where the enemy was.

Deployment

In April 2009 Lieutenant Colonel Morgan's squadron deployed to Kandahar Afghanistan, a deployment accelerated by seven months because of President Barack Obama's announcement of his troop surge. Kandahar Province is in southern Afghanistan on the border with Pakistan. To the west is Helmand Province, and to the east is Zabul Province. The climate is semiarid. Kandahar is a province of striking contrasts: most of it is a large flat desert, but on its western border, sixty miles from the Pakistani border, is the Arghandab River valley. Here, an alluvial floodplain spreads one mile on either side of the river supporting dense pomegranate orchards and trellises full of grapes. Two years later in Nashville at the Army Aviation Association of America's annual Professional Forum and Exposition, Lieutenant Colonel Morgan described the principal human settlement patterns and road networks of the region: "The environment in Kandahar was perfect for an armed reconnaissance task force. The terrain is characterized by the urban sprawl of Kandahar, the outlying villages connecting the nomadic indigenous population of Afghanistan, and the high mountain desert expanses. . . . The villages surrounding Kandahar are characterized by interconnected primary and secondary dirt roads, deep vegetated wadi systems (dry riverbeds), and mud one-story structures in the most populated areas."[2]

It was ideal camouflage for enemy insurgents. During the fighting season, the Arghandab River valley was one of the most violent places in Afghanistan. Before the troop surge, the U.S. ground force was a mere battalion. It needed a much larger force.

In the air, force levels were not much better. Tony's squadron was part of Task Force Saber, which was responsible for around-the-clock support of the ground forces from Kandahar to the Helmand River valley. To that end, Task Force Saber had twenty-four OH-58 Kiowa Warriors, nine Apaches, and two direct-support Black Hawks as well as a Pathfinder company from the 1st Squadron, 17th Cavalry Regiment, 82nd Combat Aviation Brigade, 82nd Airborne Division. With this small force, Task Force Saber had to cover all of Kandahar Province. It was a stretch. In order to cover the large expanse, Task Force Saber kept four Kiowa Warriors, two teams of two helicopters, in the air at all times. Their presence mirrored that of beat cops on patrol in a big city.

Tony's squadron was based at Kandahar Air Base, located ten miles south of Kandahar City, one of the oldest and most strategically important cities in southern Afghanistan. The air base was the largest American and NATO facility in southern Afghanistan, and in 2009 when Tony arrived, it was home to 20,000 U.S. and allied military and civilian personnel. The air base was so large not only because of its importance to the International Security Assistance Force mission in Afghanistan but also because it was supporting U.S. and coalition air operations in Iraq, part of Operation Enduring Freedom. Viewed from the air, the

Kandahar base looked like a giant automatic pistol, with the runways forming the barrel and with the accommodation, ammunition, and fuel depots and the administrative hangars forming the hand grip. The base even had a civilian airport designed and built by the United States during the Dwight D. Eisenhower and John F. Kennedy administrations.

Surrounded by interconnected concrete barriers and a razor-wire fence, Kandahar Air Base was situated on a bleached-dry light brown desert plain. During the summer fighting season, daytime temperatures were over 100 degrees Fahrenheit. The heat was only one part of the discomfort. During the day, the base was buffeted by desert winds carrying millions of tiny particles of fine sand that found their way into everything from aircraft engines to food. Tony's quarters were Spartan: at first dusty tents followed later by hastily built small plywood rooms with a bed and a shelf for books, DVDs, and CDs supplemented by some storage boxes. Embedded in the wall was a portable air conditioner/heater. There were some creature comforts to be found in the "Boardwalk," including a coffee shop and a makeshift TGI Fridays.

The mission for Tony's squadron was to set the conditions for the ground forces in the area between Kandahar and the strategic Helmand Valley. An additional 21,000 U.S. ground forces had been dispatched in March 2009 and were scheduled to arrive behind Tony's squadron.

In any conflict, the nature and pattern of fighting is dictated by the terrain. For Lieutenant Colonel Morgan, Tony, and the rest of the squadron, getting to know their area of operations and build their reconnaissance skills was of the highest importance. They began by putting large blank paper map sheets on their team tables at Kandahar Air Base. Then they drew the entire operating area by hand: the location of roads, bridges, and principal physical features as well as where the friendly and hostile forces were. The squadron took the time to map the terrain out carefully and gather aerial imagery intelligence so they could anticipate where conflicts were likely to take place and could prepare the best way to help the troops on the ground. This was vital because it enabled Tony and the Kiowa pilots to visualize where the ground forces were and put themselves in their shoes. Soldiers under fire would start to describe a particular intersection or building, and, because of his intense study, Tony usually knew exactly where they were talking about. He even knew the nicknames given by the soldiers to particular buildings. This level of knowledge gave Tony and all the other Kiowa pilots an immediate intuitive grasp of the terrain, short-ened their response time to help troops under fire by up to two to three minutes, and made their close air support far more effective. Pilots who had not done this level of study would fly to the grid and then begin searching for the location of the U.S. troops in contact with the enemy. However, they were too slow to be truly effective.

Flying over the Arghandab River valley at low altitude before the summer's lush green foliage returned and the fighting season began gave the squadron an advantage. They could see and begin to memorize every building and every walking trail where the enemy would hide and move. They called them "ratlines," and now they knew where they were.

Under Lieutenant Colonel Morgan's leadership, the squadron's professionalism was also evident in the meticulous quality of the pilots' after-action analysis. After every mission, the pilots sat in their team room dissecting what they had learned so they could pass it on to the next shift. They began by reviewing each of their knee pads, little paper notepads strapped to their knee during flight. On these little pads, the pilots wrote down in shorthand every grid in which they had a friendly or enemy interaction, the times they engaged the enemy, and how many munitions they had used. If anybody missed something, one of the others would have picked it up. Then they wrote on white boards, wrote up classified concept of operations reports, and reviewed photos of everything they had done. Next, usually with the S2 intelligence officer, they pulled up a computer and used the classified version of Google Earth to identify the exact ground over which they had flown. Unlike the Apache, which had advanced technology, the Kiowa had no electronic system to ensure that the crew had an exact grid coordinate of every munition fired. Tony recalls that in so many of these meetings, he would say "Here is the grid the ground forces gave me." The S2 would pull it up, and "I would say 'that's close, but we're actually talking about this intersection over here.'"

The Kiowa pilots would then pinpoint where the enemy had been, where exactly they had engaged them, what altitudes and what angles were involved, and what battle damage they had inflicted. As Lieutenant Colonel Morgan's adjutant, every morning Tony received this detailed information as well as the assessments from the night before. The depth of study required was comparable to graduate-level courses in an elite American university. During Tony's first deployment to Afghanistan, the level of study was as intense as the level of fighting.

In Kandahar Province during the 2009 fighting season, the pattern of warfare was unusual. Throughout military history, daybreak was always a period of high risk. But that was not true of Kandahar and most of Afghanistan. The Taliban preferred to wake up, have breakfast, and only then begin fighting, usually between nine and eleven o'clock in the morning. The Taliban loved to fight before it got dark. Once darkness fell, the Taliban broke off, ate dinner, and rested. Although the squadron flew patrols throughout the day, it was on high alert to cover midmorning and the period just before dusk. As a veteran of the Afghanistan War, Lieutenant Colonel Morgan understood the enemy's daily rhythm. Therefore, Tony's team was scheduled to take off at 10 a.m. for test fire and fly missions from 11 a.m. to 7 p.m. However, because of intense evening fighting, a flight mission would often be extended, Tony's longest lasting over ten hours.

First Air Medal for Valor

June 2009 in the Arghandab River valley was hot and humid. The paradox of the Arghandab Valley was that it was so beautiful, with lush green foliage, but so deadly. The valley was the middle of a Taliban stronghold and the center of heavy fighting that grew more intense as the summer fighting season wore on. By this time, the squadron was in at least one heavy firefight every day. That morning Tony was flying lead with CW2 David Ginn as his trail. Their mission was to support ground forces that had taken casualties. For the Kiowas, the engagement started with harassing fire, but very quickly the action became more intense. Enemy fire shifted from ground to ground to ground to air. The Taliban were firing multiple heavy weapons: Dushkas and RPK and PKM heavy machine guns.

When the ground forces radioed for a medevac to evacuate their wounded, an Air Force medevac Chinook helicopter turned up and began to take heavy damage from the Taliban's antiaircraft weapons. Tony and his trail pilot, Gunn, immediately dove into the Taliban's fierce antiaircraft fire and remained on station providing covering fire until all the wounded soldiers were safely evacuated. Had Tony and Gunn not done so, the Air Force would not have authorized its medevac helicopter to remain in the area, and the wounded soldiers would have died. The intense fighting of this engagement, as with many of Tony's engagements, left a fierce impression on the Kiowa pilots. Tragically, Ginn suffered from PTSD and took his own life in February 2021. Lieutenant Colonel Morgan, Tony, and many members of Task Force Saber attended the funeral to show the support for their fallen brother.

First Distinguished Flying Cross

August 24, 2009, Kandahar City Air Base

It was a hot, hazy day. The temperature hovered at just over 100 degrees Fahrenheit. Afghans burning trash in nearby Kandahar City helped transform the haze into smog. The winds were full of pollutants, the heat was stifling, and the atmosphere was suffocating. Technically, the mission for the 1st Squadron, 17th Cavalry Regiment, was armored reconnaissance. In reality, the squadron was an airborne rapid-reaction force, and the mission was embraced with enthusiasm. Throughout the summer of 2009 fighting was intense, and some U.S., British, and Canadian helicopters were shot down by the Taliban. The squadron knew that every day there would be many Army units calling for help.

That morning, Lieutenant Colonel Morgan and Tony received their intelligence briefing before taking off just before 10 a.m. They learned that a platoon from the 4th Engineer Battalion would be doing a route clearance patrol along Highway 1. This was a high-priority mission because the highway was littered with IEDs. Unless they were cleared, Army Humvees or Strykers could be destroyed or disabled and the road link between Kandahar City and Helmand Province would be severed. The engineers had mine-resistant MRAPs. Tony and his fellow Kiowa crews had been able to identify some IEDs from the air, but to clear Highway 1 effectively would require engineers on the ground painstakingly checking for disturbed earth or command wires. Protecting them was vital.

Before going on patrol, Lieutenant Colonel Morgan and Tony took off and test-fired three hundred rounds of .50-caliber ammunition and a full pod of seven rockets. At that time, this kind of intense daily test-firing was unheard of because Kiowas were not expected to be extremely accurate. Getting rockets and .50-caliber rounds into a predesignated target box was deemed accurate enough, but Morgan would have none of it. He had persuaded his superiors that they needed the extra ammunition for his daily test-firing in order to increase the squadron's level of accuracy. Morgan believed that combat close air support of the infantry required pinpoint accuracy: if your rocket or .50-caliber round did not hit what it needed to hit, American soldiers could die, and ammunition would be wasted. To him, this was an unacceptable pilot failure. As a result of Morgan's resolute professionalism, the squadron refined their shooting skills and improved their accuracy to levels not previously seen in the Kiowa community. Later that day, that precision would save soldiers' lives.

At 10 a.m. the two Kiowas took off into the smog from the Kandahar airfield to refuel and rearm at a forward arming and refueling point alongside Highway 1. Lieutenant Colonel Morgan had insisted that this facility be established because he believed there would be intense fighting in the area that spring and summer and that his squadron needed quick access to fuel and ammunition. Having this forward refuel and rearm base also allowed the squadron to fly with less fuel and more ammunition. Morgan's foresight would pay handsome dividends that day.

Immediately after taking off, the two Kiowas saw the engineer platoon moving along Highway 1 and relieved the Kiowa team that had been covering them. The two departing Kiowas told Lieutenant Colonel Morgan that they had received sporadic unaimed small-arms fire but nothing beyond that. Morgan and Tony assumed control and began their initial reconnaissance.

Almost immediately the engineer platoon told Lieutenant Colonel Morgan that they had identified an IED just east of Howz-e-Madad and a command wire that was heading south. The engineer platoon had recently deployed to Kandahar. It was a unit full of young soldiers with energy, spirit, and initiative. They were focused and determined to successfully execute their vital mission. In their commendable zeal, however, the engineer platoon appears not to have understood how dangerous the area south of Highway 1 actually was.

Howz-e-Madad was a Taliban stronghold with a history of small-arms attacks on U.S. and coalition forces as well as placing IEDs on Highway 1 and its side roads. There had been battalion-level air assaults into villages in the area that had run into heavy resistance.

At this point, the engineers had dismounted to follow the IED command wire. Their four armored vehicles drove slowly down the road, flanking them on both sides. For whatever reason, the engineers followed the wire a bit more quickly than Lieutenant Colonel Morgan thought wise. This was all the more dangerous because the engineer platoon did not have the firepower to fight their way out should the need to arise. Following the wire, the dismounted engineers moved about one hundred meters south of the road. Nearly twenty Taliban insurgents were lurking in the dense green foliage. They were too smart to let an opportunity like this go. The Taliban called for reinforcements; nearly eighty fighters responded. As Morgan later recalled, the insurgents were "using interconnected, preprepared, concealed, and covered positions by leveraging the wadi systems, mud buildings, and dense tree coverage." And they were well armed with AK-47s, rocket-propelled grenades (RPGs), and PKM machine guns. Some Taliban insurgents were firing from only twenty meters from the lead MRAP and one hundred meters from the engineers' fire support position on the roof of a "great putt," or abandoned mud hut. The engineers were a prime target and had walked themselves into a serious ambush.

The terrain couldn't have been better for an ambush or worse for the engineers and Kiowas. Surrounding the engineers were trees up to sixty feet tall with full summer foliage, creating an almost impenetrable blanket of lush green leaves camouflaging the Taliban below. Beneath this leafy canopy there were hundreds of trellises full of grapes, ideal hiding places for antiaircraft weapons. Scattered throughout the trees were grape huts in which the Taliban had stored caches of weapons and explosives and placed heavier antiaircraft artillery. The Taliban had covered the space where the roof had been with thick straw. When they saw the Kiowas, they removed the straw cover and opened fire.

Because the squadron had deployed nine months ahead of the bulk of the Obama ground forces surge, Lieutenant Colonel Morgan, Tony, and the rest of the squadron had become accustomed to flying over this terrain seeing Afghans walking around with Russian-made AK-47s underneath their robes. They had also learned that the Taliban had brought in Soviet-era Dushkas. The Dshk1938, or Dushka, was an old Soviet-era design but was still a lethal weapon against low-flying aircraft and helicopters. Originally developed for the Red Army during World War II as an antiaircraft and infantry-support weapon, the Dushka could fire up to six 12.7-millimeter rounds per minute with an effective range of approximately 1,100 feet. Some experts have estimated that the Dushka's rounds can penetrate 15-millimeter armor at 500 feet. The Dushkas were therefore a lethal threat to the lightly armored Kiowas.[3]

Worse, Lieutenant Colonel Morgan had learned that the Taliban had moved heavier antiaircraft weapons into the area. The U.S. pilots had seen them briefly from the air but had been unable to destroy them. The Kiowa team was looking for Russian-made 14.7-millimeter antiaircraft guns that had been emplaced in the thick-walled earthen huts camouflaged by a straw roof. The Taliban had been moving them around and as a result were hard to find.

Looking down from his Kiowa, Tony was not sure how bad the ambush the engineers had walked into was, but given that Howz-e-Madad was a Taliban stronghold in the middle of the Afghan fighting season, he was worried that this engineer platoon was undermanned and underequipped for the fight they were about to encounter. He was right.

The Taliban opened fire on the engineers with AK-47s and PKM machine guns. The engineers fired back. Over the radio, Tony could hear the enemy rounds bouncing off the MRAP's armor. Lieutenant Colonel Morgan responded with aggressive reconnaissance: flying low, banking left and right at high speed, making a lot of noise, and creating heavy rotor wash. The local Afghans called this maneuver the "angry beasts." The Kiowa team's goals were to not only identify potential targets visually but also clear the area of innocent civilians. Only the Taliban would remain. To them, the Kiowas' message was clear: if you continue to attack our ground troops, we will strike.

At first, it was difficult for the Kiowas to identify the location of the enemy fire. They asked for and received information by radio from the engineers. Over the radio, Tony could hear the urgency and stress in the engineers' voices.

Meanwhile, on the ground the dismounted engineers had dispersed into two grape huts. The first group was returning fire from behind the wall of a great putt, and the second group had climbed onto the roof of another putt to provide support fire. The engineers' CO protected his young soldiers by positioning his armored vehicles between the enemy fire and the dismounted engineers. What he did not know was that he was moving deeper into a preplanned ambush.

Once the engineers began taking enemy fire, the Kiowas were legally authorized to fire in their defense. Their first pass was dramatic: they flew in low at treetop level, then bumped up to between 150 and 200 feet and turned around and fired their .50-caliber machine gun in the same direction the engineers were firing. This enabled the Kiowas to use the engineers' fire to help them aim. The engineers could see that the Taliban fire was coming from a number of places that roughly resembled a triangle. The heaviest enemy fire came from the point of the triangle, and that is where the engineers concentrated their return fire, as did Lieutenant Colonel Morgan and his trailing pilot, Tony.

Lieutenant Colonel Morgan and Tony flew in at treetop level over the group of engineers closest to Kandahar and began firing their .50-caliber machine guns at the top of the triangle. They wanted to save their rockets for heavier weapons. From the air, Tony could see the Taliban running in different directions armed with AK-47s but had not yet seen any heavier weapons, much less antiaircraft artillery. When they made their initial pass and began firing, the Taliban watched their tactics carefully. They learned that the Kiowas were most vulnerable immediately as they broke away after firing. Because Morgan understood this threat and the terrain, he ordered Tony to change his breakaway turn based on the pattern of fire from the Taliban.

As Lieutenant Colonel Morgan made his breakaway turn from the first pass, Tony, trailing just behind, could see fire from five to six different locations within the trees. The Taliban were aiming at the underside of Morgan's Kiowa. The intense Taliban fire began ripping off tree limbs and even cutting some trees apart, clear evidence of heavy weapons. Trailing Morgan, Tony could see the trees break apart differently depending on the caliber of the weapon used. There were RPGs as well as Dushkas.

As Lieutenant Colonel Morgan's trailing pilot, Tony's job was to identify and call out the source of the enemy fire as well as fire his .50-caliber machine gun to suppress it. As Tony saw from his cockpit, the Taliban were now firing from just about every point on the compass. His short radio calls to base reported that he and Morgan were in an intense firefight and needed backup.

The two pilots flew back around and asked the engineers for help in identifying the precise sources of the hostile fire. On this second pass, Lieutenant Colonel Morgan ordered rocket fire to try to intimidate the Taliban, sending a message that this was a fight the Kiowas were prepared for. In truth, they weren't. The Taliban had successfully ambushed not only the young engineers but also the Kiowas that they knew would come to support the engineers. Later, Tony reflected that "even we were outgunned, based on the amount of force and fire that was on the ground."

This dangerous situation called for imaginative tactics. As they flew on their initial loop-out, Lieutenant Colonel Morgan briefed Tony over the radio as to how they were going to continue to engage the Taliban for the rest of the fight. They were going to change their breaks each time, executing some of the maneuvers they had practiced: bumping and at the same time changing who was the lead aircraft so that the Taliban wasn't sure which helicopter was coming in first. Morgan and Tony had also practiced cloverleafs, which meant that instead of flying the same circular traffic pattern, they were going to move around the Taliban's fire to try to get better angles to force them to change the direction of their weapons. The theory behind this was that the Taliban gunners firing from inside a great putt or deep in a great "breau" might only be in position to fire in one direction. Typically, they were positioned to fire into Highway 1. So, if the Kiowas could get south of them and force them to move or wait until they flew over them, Tony and his CO would have the opportunity to hit them from behind.

The two airmen started to change their tactics immediately. They flew cloverleafs and made different breaks. Sometimes Lieutenant Colonel Morgan broke left and Tony broke right, and sometimes they reversed that order. Their goal was to be unpredictable. They also changed their altitude, buzzing the treetops between fifty and two hundred feet. They had learned that flying between three hundred to seven hundred feet was extremely dangerous because the Kiowas were silhouetted against the sky for too long. As a practical matter, the choice was between flying very low or flying above one thousand feet. Flying low was good because it forced the Taliban to track the Kiowas between trees and buildings. Flying above one thousand feet was also good because at that altitude small-arms fire was ineffective. In this case because of the lush leaf cover, Morgan decided it was best to stay low between fifty and two hundred feet.

In twenty minutes, the Kiowas had almost run out of ammunition. On the ground, the engineers were under heavy fire. The southernmost armored vehicle was within fifty meters of a machine-gun nest and was taking heavy nonstop fire. Before the Kiowas flew off to rearm and refuel, Lieutenant Colonel Morgan decided they would destroy the machine-gun nest with their last rockets so the armored vehicle could escape back to Highway 1. Morgan and Tony fired their rockets with great accuracy and destroyed the Taliban machine-gun nest. The armored vehicle and its engineers escaped.

For the Kiowas, the next priority was to refuel and rearm. Because of Lieutenant Colonel Morgan's foresight, their forward arming and refueling point was only three to four minutes' flying time away. Thanks to the ground crews, the two Kiowas were back in the fight in twelve minutes. Morgan had trained and motivated his ground crews to what he laughingly called "NASCAR-level efficiency." No sooner did the Kiowas land than their ground crews were all over the helicopter refueling and rearming them. Tony got out of his cockpit to help his ground crew. Morgan's policy was that pilots also helped load their own rockets. Partly this was about speed. A few minutes more or less on the ground could make the difference between life and death for soldiers under enemy attack. Partly this was about squadron morale and showed the ground crews that the pilots were not above them and were willing to muck in to get this vital job done.

Tony jumped out of his Kiowa to load his own rockets and check for battle damage. He and his copilot carefully inspected their helicopter. Finding no serious damage, Tony loaded his rockets as quickly as he could. He took on only three hundred pounds of gasoline—an hour's worth of fuel—so he could maximize his ammunition load and took off.

Once Tony and Lieutenant Colonel Morgan were over the site of the ambush, they saw what appeared to be a lull in the fighting. Ever smart, the Taliban were working to confuse the American airmen by repositioning their ground troops and heavy weapons as quickly as possible. Back on station, the first thing Morgan and Tony did was to assess the problem facing the engineers who had taken heavy fire.

The soldiers were very young, and for most this was their first battle. A young inexperienced soldier led the isolated engineers over watch position, cut off from the rest of his unit under intense heavy fire. The platoon's officer and senior NCOs were trapped in their armored vehicles. So, having destroyed the heavy machine-gun nest on their first pass, Lieutenant Colonel Morgan and Tony wanted to see if the lead MRAP could turn itself around and allow the platoons' two other MRAPs to turn around and move back north. The engineers said they couldn't turn around because of the deep wadis (dry ditches) on either side of the road. They were very reluctant about attempting a K turn in an MRAP without a ground guide. Morgan agreed. At this point, deploying a ground guide into intense enemy fire meant certain death for the soldier assigned to the task, and that was something the engineers' CO understandably did not want to do.

But at the same time, the truth was that a ground guide was the only way to get the MRAP out of there. A quick aerial reconnaissance showed that the Taliban had not moved any new heavy weapons into the immediate area around the kill zone, so Lieutenant Colonel Morgan encouraged the engineers to try to start a K turn with the MRAP without a ground guide. They did but immediately came under heavy enemy fire. The dismounted engineers returned fire and were also under intense enemy counterfire. It was now clear that the Taliban were still there in full force with their heavy weapons and were determined to fight it out. At this point, Morgan and Tony started to methodically identify targets sweeping from the kill zone around the trapped MRAP and moving back east. They killed as many Taliban as they could.

As they engaged the Taliban, Tony heard an ominous sound, deeper than the Dushka and with a higher rate of fire. It turned out to be a ZSU1, a Soviet-built antiaircraft heavy machine gun. It fired 14.5-millimeter armor-piercing rounds with an effective range of between five thousand and six thousand feet. In the Soviet war in Afghanistan in the 1980s, the mujahideen had captured a number of ZSU1s and used them to shoot down Soviet helicopters. The ZSU1 now posed a lethal threat to the Kiowas. With his vast combat experience, Lieutenant Colonel Morgan immediately identified the ZSU1 as their primary target and radioed Tony: "Okay, that's our guy. We need to find it."

Finding the ZSU1 was difficult. The Taliban had a lot of different ways of moving it around. Mostly, they would attach it to a truck, drive it around, and then get it into position. Usually, that position was inside a great putt firing through a hole in the wall or through an open roof. As Lieutenant Colonel Morgan and Tony attacked other Taliban threats—dozens of armed insurgents and several Dushkas—they looked for the ZSU1.

At that point they were still on their own, as were the engineers. Enemy fire intensified. By now Lieutenant Colonel Morgan and Tony found themselves at the heart of the biggest battle in southern Afghanistan that day. But help was on the way: an airborne quick-reaction force (QRF) consisting of two more Kiowa teams and an Apache supplemented by a ground force of U.S. troops supported by French Leopard tanks. For Morgan and Tony, their job now was to buy time. They continued to support the besieged engineers but tried to use their dwindling ammunition as sparingly as possible so as to provide at least limited suppressive fire.

Unfortunately, despite their best efforts, the engineers' lead MRAP still could not turn around. Worse, Lieutenant Colonel Morgan and Tony could see that the Taliban were gradually encircling all of the engineers' MRAPs and would soon destroy them and kill or capture U.S. soldiers. Morgan and Tony knew that they had to buy the engineers time to turn their besieged lead MRAP around so they could get out of harm's way. To protect them, the airmen started to focus all of their fire in a semicircle around the armored vehicles. At the same time, Morgan told the engineers that when he and Tony had to leave to rearm and refuel, they could not stay where they were. The Taliban grasped that the only thing keeping them at bay at this point were the two Kiowas that would have to leave to refuel and rearm at some point in the ongoing battle. So, Morgan encouraged the engineers' CO to figure out a way to turn that vehicle around. If they needed a ground guide, they had to get one out there, turn the vehicle, and escape the ambush before they were overrun. The engineer officer agreed, and as his ground guide got out of the vehicle, Morgan and Tony used the last of their ammunition to lay down suppressing fire. Taking advantage of this, the ground guide helped turn the MRAP around through two points of the K turn. But when the Kiowas ran out of ammunition, the intense enemy fire resumed. Morgan asked the engineers' CO whether they could finish the turn without the guide; the CO said, "No." They needed a few more minutes of suppressive fire to get the MRAP turned around.

Lieutenant Colonel Morgan radioed Tony that they had to buy the engineers more time, and the only weapon they had left were their M4 carbines. The two Kiowa pilots had used their M4s before: if they had a small soft target, they fired their M4s rather than their rockets or .50-caliber machine gun. This situation was the opposite. The two Kiowas were facing a hard target with nothing left except their personal M4s and were taking intense enemy fire from everything from small arms to heavy antiaircraft weapons. They were also low on fuel, but if they flew off to refuel and rearm, the engineers would be overrun, and many of them would die. Tony later recalled the solution Morgan came up with:

> The idea was to make a slow pass that would draw the fire from the ground at us rather than at the vehicle and the person who was ground-guiding the vehicle and allow them just enough time to turn. So we ended up making two east to west passes right in the teeth of where they were firing from, in the direction of where all their weapons were emplaced, and it did exactly what it was designed to do: it drew the fire towards us; it did allow the engineers another two to three minutes, which was enough to get the first vehicle turned around.

Once that first MRAP was able to turn, the engineer platoon's two other vehicles were able to turn and come back too. After two passes with their M4s, Lieutenant Colonel Morgan and Tony had enabled the engineers' platoon to move north away from the scene of the ambush.

They broke off and flew to the forward arming and refueling point to refuel and rearm. Once there, they jumped out and checked their Kiowas for battle damage. Miraculously given the amount of heavy ordnance fired at the two Kiowas, there was only limited battle damage.

Refueled and rearmed, Lieutenant Colonel Morgan and Tony took off again for the third time that day to return to the fight to take out as many Taliban insurgents and their heavy weapons as possible. En route, they learned that the airborne QRF was on its way: two more Kiowa teams and an Apache attack helicopter. The Apache is an awesome fighting machine, arguably the finest attack helicopter ever built. Essentially it is a flying tank usually armed with eight Hellfire missiles, two hydra rocket launchers carrying thirty-eight 2.7-inch aerial rockets, and one M230 30-millimeter automatic cannon that can fire 600–650 rounds per minute. The Apache's advanced sensor enables it to identify and engage enemy targets night or day in any conditions.

Now the balance of firepower had changed. Lieutenant Colonel Morgan gave the airborne QRF a concise situation update: where the friendly forces were, where the enemy was, and the different routes in and out of the battle zone. When the ground QRF was about ten to fifteen minutes out, Morgan called out all the enemy positions as well as their likely escape routes.

Lieutenant Colonel Morgan marshaled his airborne force with consummate skill, directing the Kiowas to attack the Taliban from different directions. As the Kiowas attacked the Dushkas and the insurgents, Morgan asked the Apache crew to use their advanced sensors to identify and destroy the ZSU1, which they did with two Hellfire missiles. The Apache also made several passes using its cannon to cover the Kiowas. By this point, the six-hour battle had finally died down. The engineers were safely on their way back to Highway 1, the ground QRF had arrived, and U.S. troops and French Leopard tanks had begun establishing checkpoints east and west of where the battle had been to capture or kill Taliban insurgents as they tried to escape to fight another day. Morgan ensured that the ground and air QRFs were fully briefed on the engagement so they could cover the area as he and Tony flew back to their base. Later with characteristic humility, Tony summed up the six-hour firefight for me. "I'd say we got extremely lucky that day, that we didn't lose any aircraft, that we did not have any fatalities on the ground. The engineers had some injuries but nothing fatal, which was nothing more than luck on that day." The U.S. Army disagreed, awarding Morgan a Silver Star and a Distinguished Flying Cross and Tony a Distinguished Flying Cross.

A Second Distinguished Flying Cross

Christmas Eve, December 24, 2009, Howz-e-Madad, Afghanistan

The dawn was clear, calm, and cold, and the air crisp on December 24, 2008. The Taliban tended to fight when the weather was good and were very active in the spring and summer, less so in the winter. In the fall the trees lost their leaves, so the Taliban lost their camouflage. In the fall, the pilots of Task Force Saber could see the ratlines and grape rows that they used to reinforce positions. Lieutenant Colonel Morgan put a lot of pressure on S2 (intelligence) to develop an operational picture so they could understand how the Taliban were able to reinforce their positions so quickly when they were under attack. Now in the winter of 2009, the squadron could see the routes with clarity in a way they could not when they arrived in the spring and the foliage was already dense.

By December 2009, additional Obama surge forces arrived, and in Kandahar Province the United States now had a large ground force setting up FOBs and combat outposts (COPs) between Kandahar and Helmand Province. Much of the fighting centered around Howz-e-Madad, less than a mile from where the Kiowa squadron saved the engineers on August 24. A battalion level air assault to clear the area had not been particularly successful. Now in late December, Howz-e-Madad remained a vital Taliban stronghold. The Army thus decided to put two COPs on the north side of Highway 1 that looked directly down the main road in the middle of Howz-e-Madad. Snipers were in position atop both COPs.

Like most COPs, these were small fortified compounds of steel-reinforced concrete blast walls, concertina wire, and sandbags built on a base of gravel. Howz-e-Madad was located in the Arghandab Valley, one of the most violent places in Afghanistan. It was the Taliban stronghold from which the nearly one hundred Taliban insurgents had ambushed the engineers in August. The reason the COPs were built was to observe the village and to gather intelligence on Taliban movements, capabilities, and intentions. To that end, U.S. troops mounted advanced sensors and had snipers posted twenty hours a day. The Taliban resented the U.S. Army's presence but rarely ventured out as fall turned to winter. Instead, they fired AK-47s, PKM machine guns, and RPGs at the COPs around the clock.

The Christmas Eve engagement began when an officer at one of the COPs called the task force's tactical operations center to report something odd and ask them to fly over and investigate. One of his snipers on watch had spotted a Taliban insurgent walking up and down the main road in Howz-e-Madad carrying an RPG launcher. That was not unusual for the area, but what was unusual was that he kept walking up and down the road despite knowing full well that one of the snipers atop the COPs could kill him at any moment. Everyone in Howz-e-Madad was abundantly aware of the U.S. Army snipers and the threat they posed. The soldiers at the two COPs thought this an odd situation. Meanwhile, the tactical operations center was receiving signals intelligence gathered from a drone hovering over Howz-e-Madad that indicated that insurgents were massing to plant IEDs.

It was a confusing situation. The sniper thought the Taliban were going to position themselves to fire RPGs on the COP. The air tactical operations center believed that insurgents were gathering to plant IEDs. The soldiers at the two COPs did need air support but not for the reason they believed. The signals intellitence was largely correct. Taliban leaders were gathering but not to plant IEDs at the COP or to plan RPG attacks. Rather, the men with the RPGs were the bait to lure the Kiowas into a preplanned ambush.

Throughout the 2009 fighting season, the Taliban had learned not only how effective the Kiowas were in protecting U.S. and coalition ground forces but also their ability to inflict mass casualties on the insurgents. The Taliban had also learned that formidable though the Kiowas were, they had vulnerabilities that could be exploited to destroy them. In the final phase of an attack, for example, a Kiowa had no real protection against heavier antiaircraft weapons such as the ZSU1. The thin armor plates underneath the Kiowa could only protect its crew from bullets fired from AK-47s and light machine guns. Hits from Dushkas or ZSU1s could kill them and bring the Kiowa down. Another vulnerability was that the Kiowa was underpowered and not as fast as it needed to be for its mission.

Planning to exploit the vulnerabilities and destroy as many of Task Force Saber's Kiowas as possible, the Taliban set up a triangular ambush. Ironically, the site they chose for the ambush was within three hundred meters of the place where the Army engineers had been ambushed in August.

This was a strange choice, because by late December most of the natural cover that helped make the August ambush so effective was gone. The trees had lost their thick canopy of leaves, the dense pome-granate orchards were bare, and the large clusters of grapes on the trel-lises had long been picked. The Taliban were now more exposed and more easily targeted from the air.

But when the first Kiowa team arrived to answer the call from the COP, they had no idea there was an ambush. As Lieutenant Colonel Morgan and Tony flew toward Howz-e-Madad, they requested addi-tional snipers on the roof to help cover them as they flew over the town's main road trying to figure out what was going on. They thought it might be some kind of ambush, but at no point did they suspect that it would turn out to be such a large and complex one.

On their first pass, Lieutenant Colonel Morgan and Tony saw what they thought were small IEDs exploding in nearby farmers' fields. This made no sense to the airmen. There were no U.S. soldiers walking through the fields. After three passes, what the Kiowa pilots realized was that the Taliban had positioned two recoilless rifles on the east side of Howz-e-Madad to shoot down the Kiowas. They were firing at the Kiowas, but their rounds were flying over the helicopters and landing in the farmers' fields, where they exploded. The Army snipers atop the COPs helped Morgan and Tony identify that the recoilless rifle fire was coming from their left and exploding on their right. The recoilless rifle is not a weapon normally used to fire at helicopters. It is essentially a lightweight antitank weapon. Some types can be shoulder-fired, and other types are based on tripods.[4]

For Tony and Lieutenant Colonel Morgan, the immediate goal was to identify and destroy the mud hut from which the Taliban were firing the recoilless rifles. As they began to fly behind where they thought the Taliban were firing from, they discovered that the Taliban had changed their tactics. Over the previous six months, they had seen the Kiowas wreak havoc in the Howz-e-Madad area. As a result, they wanted revenge; they wanted to destroy the Kiowas.

To that end, the Taliban had set up a sophisticated triangular ambush designed to shoot down aircraft. The Howz-e-Madad Taliban had pulled in some of their top regional commanders to plan and execute the ambush. High above Howz-e Madad, a U.S. intelligence drone picked up the Taliban's cell phone and Icoms radio traffic and fed it back to S2. In near real time, Lieutenant Colonel Morgan and Tony received the number of Taliban transmissions and learned the amount of coordination that was going on.

As they maneuvered to attack the recoilless rifles, the Taliban opened up with PKM machine guns, RPG fire, and lighter machine guns. These were not their best antiaircraft weapons, but the Taliban were trying to position the two Kiowas back into the center of the triangle, where Tony and Lieutenant Colonel Morgan would face RPGs fire from roof-tops and two Dushkas from the third corner of the triangle.

The Kiowas of Task Force Saber were well prepared to deal with a triangular ambush. They understood the danger it posed and had discussed how to deal with it before deploying to Afghanistan. The key to victory was staying outside the triangle and destroying the antiaircraft weapons at each of the three points of the triangle along with their operators and support personnel. In dealing with this ambush, the Kiowas' exceptional maneuverability was complemented by the pilots' extremely high level of training, and the exceptional accuracy of their marksmanship as well as their knowledge of the terrain.

As a result, despite the heavy enemy fire, Lieutenant Colonel Morgan was reluctant to leave the area. And once he saw the complexity of the ambush and heard the Icom chatter, he was convinced that there were too many enemy weapons and too many senior Taliban commanders on the battlefield to fly away. As Tony later recalled, "What began as a target of opportunity for the Taliban became a target of opportunity for us. There was significant enemy capability we could take off the battlefield."

As they had discussed before deploying to Afghanistan, Lieutenant Colonel Morgan and Tony maneuvered sharply and began to fight their way out of the triangular ambush. The key to victory was staying outside the triangle and destroying the antiaircraft weapons at each of the three points of the triangle individually along with their operators and support personnel. To do so, the airmen needed more firepower. Morgan realized the scale and complexity of the Taliban ambush and called for assistance.

At that time in Kandahar in 2009, there were always two Kiowa teams airborne at all times. One team was patrolling Highway 1 toward Helmand Province, and another was patrolling the Arghandab Valley. Lieutenant Colonel Morgan summoned them. Later in the engagement, he summoned his QRF, a "pink team" of two Kiowas and an Apache from the Kandahar airfield.

In all, the battle between the Kiowas and the insurgents lasted six hours, with multiple refueling and rearming runs. Because Lieutenant Colonel Morgan had transformed the squadron's culture, he and the other pilots got out of their helicopters, joined the crew chiefs, and helped rearm the Kiowas. As a result, rearming and refueling the aircraft now took only fifteen minutes.

Once back on station, Lieutenant Colonel Morgan and Tony destroyed the recoilless rifles, which ended one point of the triangular ambush. They weren't the biggest or most threatening weapons, but they were the best camouflaged. After that success, the two Kiowas went to refuel and rearm. They linked up with the two Kiowas that had been patrolling the Arghandab Valley and reentered the fight as a four-aircraft formation and now focused on the Dushkas and PKM machine guns. In the final ninety minutes, the QRF from Kandahar joined in. Together, the six Kiowas destroyed not only the remaining Taliban antiaircraft weapons at the other two points of the triangle but also significant numbers of insurgents carrying RPGs. In all, Task Force Saber's Kiowas killed between eighty and one hundred insurgents (some of them senior Taliban commanders) and destroyed a lot of dangerous enemy weaponry. Many years later, Tony reflected on this intense Christmas Eve battle with characteristic humility:

> We got superlucky: no aircraft were shot down. We did take some aircraft damage, but we had achieved our goal. We had removed their heavy pieces and all of these Taliban commanders who had helped set this ambush up. This was not your average Taliban fight. For the most part they liked harassing attacks trying to lure soldiers into an ambush, like the attack on the U.S. Army engineers in August 2009. Usually, depending on the number of U.S. troops on the ground at the time, the Taliban would try to overwhelm them with sheer mass and with typical flanking maneuvers. This time was different. We only saw that level of sophistication two or three times during our deployment.

By January 2010, the pace and intensity of Tony's deployment in Kandahar Province diminished. Throughout the rest of the winter, there were only two significant engagements: one in the Argonaut Valley and one in the Helmand River valley. Other than these, Tony experienced nothing worse than sporadic harassing fire.

Within a few months of this decisive victory, U.S. forces mounted a clearing operation in the village of Howz-e-Madad. In April 2010, the 101st Airborne Brigade worked closely with the Air Force and leveled the first five hundred meters of the village south of Highway 1, which contained a collection of Taliban fighting positions. By the early summer of 2010, U.S. forces had driven the Taliban from Howz-e-Madad, for so long the Taliban's stronghold in Kandahar province.

Throughout their deployment, Task Force Saber enjoyed an enviable reputation for its members' deep commitment to the ground forces as well as their supreme professionalism in carrying out their duties. The unit's call sign was Seamus. Every member of the unit made little tabs for their uniforms that said "Seamus." Tony recalled, "I can't tell you how many times I was stopped around the Kandahar airfield by officers and soldiers cycling in or out or on resupply missions to receive their grateful thanks. We had an immense reputation at that time because everyone put that level of effort and study into their work, into being a professional pilot."

21

ROSS PIXLER

Iraq, 2007–2008

After Ranger School, Ross Pixler chose Fort Benning as his first duty station. He wanted to take the Reconnaissance Surveillance Leaders Course, the graduate level of Ranger School, but the Infantry Officers Basic Course staff would not send him because April was pregnant. Instead, since he was going to join a heavy cavalry unit, they sent him to the Bradley Leaders Course, which he passed in July 2006.

In August Ross joined the 3rd Squadron, 1st Cavalry Regiment (3-1 Cav), operating as part of the 3rd Brigade, 3rd Infantry Division. That year, the Pentagon had added cavalry squadrons to a number of infantry brigade combat teams. The goal was to bring new skill sets as well as a stronger emphasis on conducting reconnaissance surveillance of the enemy so the infantry units could attack and destroy them.

When Ross joined the 3rd Brigade in August 2006 at Fort Benning, he was excited to be not only a new platoon leader but also part of building something new. He served first in the squadron headquarters under Capt. Jimmy Hathaway, an outstanding inspirational leader. Ross then moved to Charlie Company, commanded by Capt. Ernie Melton, who was later court-martialed for stealing a .50-caliber machine gun from another company to embarrass its commander and negatively affect his evaluation.

By January 2007 Ross and his soldiers arrived at the Joint Readiness Training Center (JRTC) at Fort Polk, Louisiana. There, units rotate through for a month prior to deployment. The JRTC is the U.S. Army's leading combat training center, a rigorous testing ground for any unit before going into harm's way. The JRTC's mission is to improve each unit's effectiveness as a fighting force. To that end, its exercises simulate warfighting as realistically as possible. The exercises are not tightly scripted scenarios but instead are fluid, fast-moving battles in which the opposing force plays a smart enemy who anticipates your unit's moves and countermoves. Officers, NCOs and soldiers have to adapt quickly and think fast. No mistake goes unnoticed.

At the end of February 2007, the 3-1 Cav began deployment to Iraq. Ross flew out on the advanced echelon flight to Kuwait a week ahead of the flights carrying the rest of his cavalry troop. Those aboard the flight were the advance party responsible for putting in place the support systems to enable the rest of the unit to arrive smoothly. On March 2, the unit arrived in Kuwait. They were there for several weeks getting prepared to move into Iraq.

By this point, six months after joining the unit, Ross concluded that Captain Melton's leadership was "destructive and untrusting." Building the trust that was so essential with the troops in his platoon was very difficult because the men distrusted their officers and their leadership.

When Ross and his soldiers first arrived in Iraq in March 2007, there was a revealing incident that illustrated why the soldiers distrusted their officers. Unfortunately, Ross, a young and inexperienced officer, made the situation worse. He should have taken ownership of some ill-thought-out orders but didn't. As Ross later admitted, "I probably did not do what I should have done."

In March 2007 Ross's cavalry squadron arrived at a brand-new FOB, Hammer. His unit's initial task was to defend FOB Hammer and protect the brigade's headquarters. FOB Hammer was still being built. At that time, facilities were primitive. Ross and his soldiers had nothing but tents, portable bathrooms, and limited electricity. FOB Hammer was located on the southern part of the Butler Range Complex, formerly the Iraqi Republican Guard tank range. When Ross and the rest of his cavalry squadron arrived, FOB Hammer did not have a perimeter wall. It didn't even have concertina wire around the whole FOB. At the time, concertina wire covered only two-thirds of the perimeter. The remaining third was open for anyone, any insurgent, to walk in. At that time, there were no security towers and no fixed defensive positions.

At FOB Hammer, Ross's first job as a platoon leader was supporting base security. His first patrol took place around 3 a.m. He was ordered to do cloverleafs, a counterreconnaissance mission to see if any insurgents were getting through their lines or scouting out their position. This was an important but potentially dangerous mission. Because the FOB was located in a former Iraqi Republican Guard tank range, there was a lot of unexploded ordnance scattered around—Iraqi, not American—and nobody at FOB Hammer knew where it was. Ross remembers thinking that "this cannot be more dangerous: having to walk in the middle of the night outside the wire where it hasn't been cleared; where there is unexploded ordnance. To find something, we were being asked to walk around until we blew up." Worse, there was no natural light; there was no moonlight because of a thick cloud cover, and the soldiers' night vision devices could not operate well in the limited visibility with no ambient light. Using flashlights would enable anyone potentially scouting their positions to easily identify and shoot them. Worse still, they had no metal detectors, and their radio communications with FOB Hammer were at best intermittent.

In these circumstances, Ross thought that the mission was crazy. He remembers arguing with his troop company commander, Captain Melton, that the mission was "stupid" and that a different approach to it was necessary. It could be executed safely, Ross argued, if they had metal detectors and better night vision goggles and changed the time of the patrol. Ross added that if the patrol took place at 11 p.m. instead of

3 a.m., the moon would be out so that Ross and his soldiers could see what they were doing. Captain Melton told Ross to "shut up and color," rejected his protests, and ordered him to proceed with the mission.

Ross's soldiers also thought the mission was stupid. They told him that "this does not make sense." He made the mistake of telling them that he agreed, that he had told Captain Melton so, but told them that they were professionals and that orders were orders. Now his soldiers knew that Ross disagreed with the mission. Ross had made a mistake. As he recalled many years later, "this is not what a good leader does, so in a sense I had garnered some of their trust but in another way proved that I was untrustworthy because they knew that I knew that the mission was dumb but we were going to do it anyway. For the soldiers there were two ways of looking at it."

Captain Melton had given an order, and Ross had no alternative but to obey and was going to carry out the order in the safest way possible. To minimize the risk to his soldiers he went first, and his soldiers followed in single file walking in his footsteps with a ten-foot dispersion between each soldier. They completed the mission successfully without any casualties. To minimize future risk, Ross went out alone to look at the ground during the day. He identified safe routes that he and his soldiers could follow day or night. This helped him begin to earn the trust of his soldiers. By the end of Ross's deployment after intense fighting with enemy insurgents, this first patrol "looked like nothing" to Ross, but it was his first patrol, and to him and his soldiers it felt dangerous.

In May 2007 Ross and the rest of 3-1 Cav moved out of FOB Hammer and into COP Cashe North, seventeen miles southeast of Baghdad. Named for Sfc. Alwyne Cashe, who received a posthumous Silver Star, the COP was located inside the town of Tuwaitha, near the old Osirak nuclear reactor that the Israeli Air Force had bombed in June 1981. Another company was deployed in southern Tuwaitha. Under Saddam Hussein, Tuwaitha had been a government center, but after the Iraqi dictator's fall, much of the town was abandoned to insurgents. Wild animals had ravaged it, leaving large quantities of fecal matter. It was filthy.

It was here that Ross encountered another example of "destructive and untrusting leadership" when his immediate superiors supported an investigation into him for firing his weapon in a firefight and left him alone and unsupported when he had to report to a Judge Advocate General (JAG) Corps officer in Baghdad to account for his actions.

Here is what happened. In May 2007, Ross had led a Small Kill Team attempting to intercept insurgents planting IEDs on Route Viking, one of the major thoroughfares going into Baghdad. Ross brought his team to a spot east of the Diyala River bridge, just north of Jisr Diyala. It was a classic cavalry mission: identify and report back what was happening. Ross found a good concealed location from which to conduct the surveillance and reconnaissance: a big, five-story blown-out building. Inside, there was nothing left but concrete pillars right to the top where Ross and five of his soldiers were concealed. They had clear views and excellent fields of fire if needed. Atop the building, well hidden and with binoculars, they could see everything. Right across the street, there was a big unused warehouse surrounded by a walled compound. Behind the warehouse was the main road, which Ross and his team could see clearly.

Before they went up to their observation post, Ross and his soldiers thoroughly searched all of the surrounding areas, making sure there was no one around who could give away their position to the enemy. They also searched the compound and the warehouse for explosives and hid their vehicles off to the side away from the compound.

About an hour into their Small Kill Team mission, Ross and his soldiers saw a car approaching at high speed. It screeched to a halt and then turned left. A young military-age man got out and jumped into the bushes. The car turned left down a side street and drove away. This was very odd. Ross and his soldiers could not understand why the young Iraqi man did that. At first, they thought that perhaps he was hiding in the bushes because he had seen them. But they didn't know for sure and remained concealed, focusing on their mission.

About twenty minutes later, another car came barreling down the main road at high speed from the opposite direction of the first car. The second car stopped near the bushes, and the young Iraqi man who had been sitting in them suddenly jumped up and got inside the car, which drove into the walled compound right across the street from Ross and his concealed patrol.

Through their binoculars, they noticed another puzzling and unexpected development: cars and trucks were not traveling on the main road as they normally would. Instead, they were traveling on the side roads immediately adjacent to the main thoroughfare.

When the suspicious car turned into the walled compound, five military-age Iraqi men got out. They began pulling out weapons, ammunition, and other military equipment. To Ross, it looked like they were setting up their own overwatch post atop the walled compound right across the street and appeared to be planning an ambush against any U.S. or coalition forces that might travel up or down Route Viking. Ross tried to report this in but couldn't reach company headquarters because of poor communications.

Suddenly, Ross noticed that the car engine had started up and that two of the five insurgents now looked like they were getting ready to leave. To stop them, Ross ordered two of his soldiers to go downstairs, get in their Humvees, and prevent the insurgents from leaving the compound so they could be captured. He also ordered the specialist next to him to fire two warning shots right in front of the car containing the insurgents to get them to stop. The specialist's two shots were perfect, hitting the dirt just in front of the insurgents' vehicle and bouncing up into the surrounding mud wall. And it worked. The car carrying the insurgents slammed on its brakes and backed up into the compound.

The three insurgents not in the car began shooting at Ross and his soldiers on the rooftop. For a time, the firefight was intense. Around eighteen or nineteen rounds hit the floor and the low wall within inches of Ross. Later he recalled, "I don't know how I didn't get shot in all of this, but by the grace of God, I didn't. So many rounds whizzed past me." Ross ordered his soldiers to return fire. They killed two of the insurgents immediately. The other three retreated back into the walled compound. They then tried to escape, with two running out to Ross's right side and one to his left. Ross told his soldiers to engage.

The specialist on Ross's right said that he and the others couldn't fire because the insurgents had dropped their weapons. Ross replied, "I don't care if they have dropped their weapons. Shoot them!" He did not have time to remind his soldiers of the nuances of the laws of war: you don't become a noncombatant by dropping your weapon. You become a noncombatant by an act of surrender. The insurgents were not surrendering; they were retreating. But because the specialist and the other soldiers did not trust their leadership, they told Ross that they couldn't shoot because if they did, they would be investigated.

Ross responded by picking up his weapon, the first time in this deployment that he had to use it. Once Ross fired his rifle, his soldiers followed suit. Ross hit the insurgent who had fled to his left in the calf and was able to capture him later. Immediately the insurgent went down, and Ross switched his attention to the other two insurgents and opened fire. It was a difficult shot. By this point, the two insurgents were four hundred meters away and running hard. And Ross had no experience firing down at a high angle from a five-story building. He missed narrowly, as did his soldiers, but the insurgents knew that sooner or later they would be killed or wounded and decided to surrender. Once they stopped and raised their hands, Ross told his soldiers to cease firing. They came down and captured the two insurgents. Next, they captured the wounded insurgent. In the end, they killed two insurgents and captured three.

When Ross reported the engagement to his immediate chain of command, he was put under investigation. Eventually they sent Ross to see a senior JAG officer in Baghdad to explain and justify his actions. The JAG officer was a Navy captain who was astonished to discover that Ross was under investigation. Not only did he agree wholeheartedly with the way in which Ross handled the incident, but he also had Ross explain to his staff the distinction between a combatant who drops his weapon in flight and one who formally surrenders.

What is disturbing is the way in which Ross's immediate chain of command abandoned him, leaving him to fend for himself. Later Ross recalled, "I was sent to Baghdad with no leadership supporting me whatsoever. In fact, it was quite the opposite: they were the ones encouraging the investigation of me." Ross was not surprised when he later heard the news that his company commander, Capt. Ernie Melton, was investigated and court-martialed for stealing a .50-caliber machine gun from another company's inventory to help ruin that company commander's career.

Because COP Cashe North was located in Tuwaitha, the U.S. Army built it as securely as possible. Around the perimeter was an earthen berm, one hundred feet high, made of earth, rocks, and stones. Atop the berm were a series of guard posts manned twenty-four hours a day. The entrance was cut into the berm in the shape of a dog's leg so that the insurgents could not drive car bombs directly into the middle of the camp. The buildings inside the square camp had concrete floors and solid brick and masonry walls. For Ross and his soldiers, COP Cashe North was fairly secure from direct fire, but they remained vulnerable to indirect fire from insurgents firing mortars.

The move to COP Cashe North brought a change of mission. Here, 3-1 Cav supported the main effort to their south fought by the 1st Battalion, 15th Infantry Regiment, 3rd Heavy Brigade Combat Team, 3rd Infantry Division. They were on the front line; Ross's company was between them and Baghdad. The mission was to interdict the flow of accelerants (explosives and insurgents) flowing into Baghdad down Route Six, the main road into Baghdad. Another part of the mission also included stopping insurgents coming from the south and from Iran. Complicating 3-1 Cav's operations was that the soldiers were located on ethnic fault lines between pockets of Sunnis and Shiites who were constantly fighting each other as well as Ross's company. He recalled, "You never quite knew who you were talking to, much less what their agenda was."

Those conversations were inhibited because Ross and his platoon often did not have an interpreter. In June and July 2007, his platoon and the rest of 3-1 Cav did a lot of foot patrols walking around COP Cashe North through the palm groves and the neighborhoods. Usually, these were long patrols in the middle of the night and, during the day, identifying any suspicious activity. If they needed to have an intentional encounter with a local, they could borrow Captain Melton's interpreter when he was available.

By the late summer when Ross and his soldiers began patrolling out of their new COP, the fighting got a lot worse. And for the first time, they began to encounter a lot of IEDs. Several soldiers in other platoons in the company were wounded either by insurgent gunfire or IEDs.

This is when Ross lost his first soldier, SSgt. Allen Greeka from Alpena, Michigan, who died on Friday, July 13, 2007. Ross and his squad were on a twelve-hour patrol in Jisr Diyala along the banks of the Tigris. Insurgents were trying to use outlying areas around Baghdad to store caches of IEDs. Ross's mission was to look for caches of IEDs as well as elements of IEDs that insurgents might use and bring into Baghdad. The conditions were miserable: the temperature was 130 degrees with high humidity off the Tigris. They were patrolling through the palm groves in the western part of Tuwaitha outside COP Cashe North alongside the Tigris River. As Ross recalled, "We stopped for lunch at a destroyed house. We ate what we could from MREs and then set off for the remainder of our patrol into a palm grove. I was giving everyone a hard time for not being eager to get back on our job. I thought they were taking too much time eating and relaxing."

Slightly irritated, Ross stood up ordered his soldiers to finish, get in formation, and get moving. SSgt. Allen Greeka started yelling at everyone to get up. As Ross recalled, "Allen ran in front of me and said jokingly that I could not have the privilege of being number one on the patrol. He was always a jokester, truly a smartass but always in a funny way, always lighthearted." Greeka, always a dedicated professional, led the patrol out of their lunch break and pushed back into the palm grove. Ross halted the squad because SSgt. Geeka saw something suspicious in the dense vegetation that looked like it might be an IED cache point. As he later told me, "What made me suspicious was the way that the palm trees were growing. In those groves, the palm trees were usually evenly spaced but in crooked lines. But these ones were growing around a small circle in which there were some bushes and trees, and you don't find that in a palm grove. It looked out of place, and I wanted to take a closer look."

Ross moved toward the small circle of bushes to take a closer look at the suspicious area. He told his soldiers to stand still and stand back. At this point, Staff Sergeant Greeka was in front of Ross. When Ross left the formation, Greeka did what a good NCO would do: he started going down the line of soldiers seeing how each one was doing in the extreme heat, checking each soldier's water and ammunition. In between the second to last and last soldier, Greeka stepped on a crush wire and triggered an IED under a palm frond. In the explosion, he lost both legs and an arm; Ross's medic went to work attempting to save Greeka's life. Ross ran over while everyone else was ordered to pull security and assisted in applying tourniquets to stop the bleeding, but their efforts could not save Greeka. Because Ross had dispersed his squad carefully, no one else was wounded by the IED explosion.

It was a strange and terrible tragedy. Ross himself and eight members of his patrol had walked over that spot in the palm fronds and did not step on the crush wire. It was underneath a palm frond, and there were palm fronds everywhere. To this day, Ross feels the loss of Staff Sergeant Greeka very deeply and still agonizes over what he did that day. He told me, "Had I not stopped us at that moment, had I stopped the patrol twenty feet further on, perhaps Allen would not have gotten killed." Perhaps, but Ross did not stop where he did carelessly or impetuously. He stopped where he did because in his best professional judgment, there was a potential threat to his and other soldiers that had to be investigated.

As the summer of 2007 wore on, the fighting intensified further. The company lost several Bradley fighting vehicles, including two of Ross's. A number of soldiers were wounded. One of them was Private Innes. Originally from Jamaica, Private Innes was well liked and well respected. He was trapped inside a burning Bradley after it had driven over an IED, part of a double IED hit on two Bradleys. The Bradley commander and the gunner were able to get out but could not get

Private Innes out through the driver's hatch because the impact of the IED had warped the metal and trapped his leg inside the vehicle. Worse, the Bradley caught fire, and the high-explosive and antipersonnel ammunition inside began to "cook off," sending bullets ricocheting throughout the interior of the cabin.

The only way to rescue Private Innes was for soldiers to open up the back ramp of the Bradley, crawl in past the flames and through the narrow "hellhole" located between the turret and outer exterior of the Bradley, pull his chair backwards and downward, and pull him out through the flames and rounds. Two soldiers put their own lives at risk braving the flames and the ricocheting ammunition to save Innes's life. In recognition of their courage, Ross nominated the two soldiers for medals for valor, but despite his best efforts, the company commander denied the nominations on the absurd grounds that the two soldiers were merely doing their jobs. To this day, Ross sees it as a personal failure that they did not get the recognition they deserved by their selfless act of courage. He is being too hard on himself. Many worthy nominations fail to make it through the bureaucratic process and poor leadership at the company level.

By August 2007, Ross was finishing up a year as a platoon leader with the 3rd Squadron, 1st Cavalry Regiment, and had to transition out. Ross learned that two lieutenant-level positions had opened up, and he and his peer (another platoon leader in Charlie Company) had two choices as to where they wanted to go. One choice was being a platoon leader in the 1-15 Infantry Division's Alpha Company (known as Hardrock Company), which had suffered the heaviest combat casualties in the entire brigade and identified as the brigade's main effort. They were on the front lines covering the southern area of the brigade's area of operations on the approaches to Baghdad. A stronghold of Sunni militias and Al-Qaeda terrorists, it was a dangerous tinderbox ready to explode into intense violence. The other choice was taking over an XO position in another troop in the same cavalry squadron in a relatively peaceful area in the middle of the brigade's area of operation. Ross got to pick first, and for him it was an easy decision. He wanted to be a platoon leader again. To Ross, going into frontline combat was more challenging and more interesting. Leading troops in combat was what he had trained for. Staying in the cavalry squadron where he had experienced such a lack of leadership was not an option. Later Ross reflected, "There were some great leaders at 3-1 Cav, but I was experiencing a lot of bad things that should not have happened. I was very thankful to be at 1-15." When Ross left 3-1 Cav for the 1-15 infantry, he moved from COP Cashe North to COP Cleary, the battalion headquarters, and then on to COP Cahill.

Ross made the right choice. His new COs, company commander Capt. William Clark and battalion commander Lt. Col. John "Jack" Marr, were exceptional leaders. Ross recalled that "both of them had trust in me, and I know that I was given authority and responsibility because they trusted me to get the job done. It was because they trusted me that I trusted them."

When Ross first joined the battalion, Lieutenant Colonel Marr wanted to assess and evaluate him, so Ross spent ten days at COP Cleary on battalion staff assisting its S3 operations officer, Cpt. Josh Powers. Ross quickly observed what an excellent commander Lieutenant Colonel Marr was, describing him as "down to earth" and "a really good listener." Ross also got excellent advice and guidance from Sgt. Maj. Mark Moore. Lieutenant Colonel Marr assigned Ross to command the 3rd Platoon, Alpha or Hardrock Company, located at COP Cahill.

In some ways, Ross's task in building trust with his new platoon was more difficult than with his old platoon. To begin with, he was taking over as the leader of a platoon in middeployment, which is never easy. The biggest difficulty, however, was that he was taking over from a lieutenant who had essentially been fired for being toxic. Ross had heard many bad stories about the lieutenant, and unfortunately they were mostly true. The young officer was arrogant and distant and thought he was too good for his NCOs and soldiers and rarely lost an opportunity to tell them so. He was someone who rarely consulted his platoon sergeant, Pete Black, a thirty-six-year-old combat veteran. Worse, the lieutenant refused to do the things he would order his subordinates to do. Worse still, he rarely listened to his soldiers' concerns, so after this experience the soldiers did not like or trust officers at all.

Ross's first step was to have a long, frank conversation with his platoon sergeant about what they expected of each other and how they were going to work together and lead the platoon as a team. Platoon Sgt. Pete Black, who had taken over the platoon in June 2005, had joined the Army as a private in 1989 and had deployed multiple times. In that conversation, he was blunt. He told Ross that "the men come first no matter what. I don't care about you. I don't care about anybody else. Never compromise the respect for and integrity of the platoon. You don't take from my men."

Sergeant Black explained that successfully accomplishing the missions assigned by Captain Clark or Lt. Col. Marr required soldiers who were well taken care of: well-equipped, well-trained, well-counseled, well-tutored, and well-developed. Sergeant Black said that what he needed was someone he could rely on when they had to run split patrols.

Ross described Sergeant Black as "incredible," an NCO who was a father figure to every soldier in the platoon and knew much about the soldiers' lives and would do anything to help them. Ross listened carefully and respectfully before setting out his own expectations. Despite the difference in age, education, and family background, Ross and Black recognized that they were both cut from the same mold: courageous, selfless leaders who cared deeply about their soldiers. The new young lieutenant and the veteran platoon sergeant began to build a special bond of trust.

Ross acted on Sergeant Black's advice, but what also helped him build trust with his soldiers was that "he was always one of the guys." As Pvt. Tony Martinez recalled, Ross "did not pull rank: he treated us as equals. And if we were outside smoking a cigarette or talking after a long mission, he was with us. He quickly established himself as a brother."

In addition, Ross led by example. No matter what task the platoon had to do—filling sandbags, laying concertina wire, or pulling security—he was doing it with them. Inside the wire, Ross volunteered for tasks so his soldiers could rest. In combat missions, if there was a door to go through, he was willing to be among the first to go in.

This was a veteran platoon that had experienced multiple deployments and suffered numerous casualties in often heavy combat. Most of the soldiers in it had been part of the initial Anglo-American invasion of Iraq in 2003 and had endured multiple combat deployments since then. One soldier had even served in the platoon for over ten years. The squad and section leaders had been privates under Sergeant Black on earlier deployments. He had "raised them" from within the platoon. They were a close-knit team who knew a courageous, selfless leader when they met one. By the end of September 2007, the soldiers had confidence in his leadership abilities and his talents.

Ross began building rapport with his soldiers on their very first patrol together. The departing platoon leader was tasked with showing Ross around the platoon's area of operations. They were driving together in a Humvee along with one section of the platoon. After going outside the wire, they paused to fire their weapons on a makeshift range in an open desert area to confirm that there was no risk of hitting innocent civilians. Unfortunately, a cow somehow wandered in and was accidentally killed by rifle fire from the platoon.

The farmer who owned the cow came running out very upset, screaming complaints about the American soldiers who had killed the animal. The departing lieutenant began writing up the farmer's complaint so the Army could compensate him for his loss. One of the soldiers noticed that there was a large IED sitting in the farmer's driveway. The departing lieutenant asked the farmer about it. He replied that it wasn't his, that insurgents had come by during the night and left it there. "What about my cow?" the farmer screamed again. The departing lieutenant returned to writing up the complaint.

Ross noticed the soldiers' concerns and quickly concluded that the farmer was actively helping the insurgents and told the outgoing lieutenant to stop writing up the farmer's complaint. Ross was in an awkward situation: he was not yet the platoon leader but decided that he had to stand up to the officer he was replacing who appeared heedless of the fact that the farmer was an active supporter of the insurgency and was holding on to the large IED until the insurgents could come back for it. Ross and the soldiers could see what the other lieutenant could not. So, Ross spoke bluntly to the other lieutenant: "Over my dead body are we paying this farmer compensation. This guy is bad." He tore up the paperwork for the compensation claim and then scolded the farmer. Ross told him that he and his soldiers were taking the large IED away and that if he continued to support the insurgency, they would come back and arrest him as well.

Ross's handling of this situation did a lot to garner trust within the platoon. Not only had they experienced an officer who listened to them, but also an officer who grasped reality on the ground and did not misread encounters with Iraqis. It was a good start. Ross soon built a strong bond with Sergeant Black, and they ran the platoon "sewn together at the hip."

Ross recalled, "I knew I had some success when they invited me to work out with them. The platoon was well known for lifting in the gym when they weren't eating or on patrol. They started having me work out with them, and when they finally jumped me, I knew I had been accepted as one of the platoon."

As Ross cemented his relationship with his soldiers, he had the support of a first-class chain of command in Capt. William Clark and Lt. Col. Jack Marr. A native of Clinton, Iowa, Captain Clark was an inspirational leader. He had brown hair worn in a crew cut, blue eyes, and chiseled features tapering into a strong, protruding jaw. He was a distinguished military graduate of Northern Illinois University's Reserve Officers' Training Corps program and received his commission as a second lieutenant in 1999. Before his captain's career course, Clark had deployed to Kosovo as part of Operation Joint Guardian, the U.S. military's peacekeeping mission there.

When Ross joined Hardrock Company, Captain Clark gave Ross his initial counseling. Part of that counseling was the importance of inspirational leadership that Ross has incorporated into his leadership philosophy to this day. Clark based his counseling on a leadership philosophy document that he had written. Even at a time when the Army was emphasizing inspirational leadership, it was most unusual to find a young captain who had not only given so much thought to leadership at the company level but had also developed a leadership philosophy set out in a well-crafted paper. Ross was deeply impressed.

Captain Clark also helped Ross develop the diplomatic skills necessary to work with the local sheikh, tribal leaders, and government officials. Ross remembers Captain Clark (who sadly took his own life in September 2018) with great admiration and respect: "He was a very quiet, soft-spoken, unassuming person. I think I saw him smile once. He was always calm, reserved, someone who did not change emotions very easily. He was extremely articulate, extremely competent, technically and tactically."

What amazed Ross and the other platoon leaders and first sergeants was Captain Clark's ability to draw together a large body of information, process it, and organize it into a clear, readily understandable mission briefing sometimes at short notice. Ross and the other platoon leaders were equally impressed with the painstaking care and clarity with which Clark developed orders so that missions would be executed efficiently and effectively.

One thing Ross noticed was that Captain Clark was one of the only company commanders who didn't have a Ranger Tab. Among other things, Ranger School taught the nuts and bolts of mission planning, but although Clark had not earned the Ranger Tab, he didn't need it. Ross added that Clark

> was passionate about being the best tactician he could be; he would include in his plans things the rest of us would not have thought about. That helped to build the cohesiveness, not only with the soldiers in his company, but with the Sons of Iraq and local Iraqis. They cared about him so much that when they got bad intel that COP Cahill had been attacked by insurgents and Captain Clark had been kidnapped, they came roaring down the road in cars and trucks loaded with guns and ammunition ready to rescue him. All they wanted was to be told where to go.

If Ross's company commander was outstanding, so was Lt. Col. Jack Marr, his battalion commander. Marr graduated from the Marion Institute, Alabama's state military college, in 1987 and was commissioned a second lieutenant in the infantry at age nineteen. He was not only a combat veteran but also a defense intellectual and a leader deeply invested in developing and nurturing young officers.

Ross had the chance to observe Lieutenant Colonel Marr's leadership style, which he admired, while serving on Marr's battalion staff. Now as a platoon commander, Ross could more fully appreciate Marr's leadership style. Armed with a few cigars, Marr visited each company regularly to meet with the company commander and the four platoon leaders. Although he shared with them his assessment of the progress of their mission, the main reason Marr was there was to listen carefully and respectfully to what each of his subordinates had to say: their assessments of conditions on the ground, what they were seeing, and what they were thinking. Ross recalled that "Lieutenant Colonel Marr wanted to know what we thought, what was happening on the ground, what was working, what was not in terms of tactics, and what changes needed to be made. He really wanted to hear our perspective."

Above all, Lieutenant Colonel Marr trusted his young officers. For Ross that was the biggest difference from 3-1 Cav. Then, when there was a shooting, an incident, or even an attack, the assumption was that somebody had made a mistake. Under Marr's leadership, that was never the case. His assumption was that his officers were always doing the right thing, that incidents, shootings, or attacks were a natural result of war and the enemy's desire to kill U.S. soldiers. Ross knew that Marr would do the right thing.

The year 2007 brought two major new developments that affected Ross, his soldiers, and all American fighting men and women. The first was that the violence on the ground became more intense. The second was the implementation of the surge and the introduction of the new counterinsurgency (COIN) doctrine, which meant that Ross and his troops left FOB Hammer and moved to a smaller COP more integrated into the community. Ross supported the doctrine advocated by the charismatic Gen. David Petraeus, the new commander in Iraq. Ross also admired Petraeus, describing him as "an amazing, great leader. What I recall was that he was not afraid to come out and see us where the fighting was fiercest. One of his helicopters was shot at when attempting to get to us. It was a hard landing, and his helicopter needed to be fixed at our COP. He also had a way of taking the COIN doctrine and applying it at the lowest level. Application is key; otherwise, doctrine is useless."

To Ross, what was so impressive was the way General Petraeus ensured that every unit from brigade to battalion to company to platoon not only knew their mission but also understood how their mission fit in with the larger unit of which they were a part. Ross explained that "everything was nested, and trust was given and received. I did not see the same level of understanding in Afghanistan."

Al Bawi is a small village near Salman Pak. In English, Al Bawi means "bow" and refers to where the Tigris River bends because of a piece of land that juts into the river. That is where Salman Pak is located. It's a town with a dark, sinister past. Under Saddam Hussein, Salman Pak was home to a large biological warfare research laboratory capable of producing two hundred quarts of anthrax per week.

At COP Cahill, the morning of October 30, 2007, was clear and cold. Ross and his platoon were based just north of Salman Pak on the string side of the bow, right off the side of the main road known as "Route Wild." They had received intelligence that suggested that there was a cache of weapons located nearby, so that morning their mission was to find and destroy it. They set out in a convoy with Ross's Bradley in the lead followed by two Humvees, with another Bradley bringing up the rear.

On their second stop that morning, they had to drive their Bradleys off the road. Intelligence had guided them to a grid coordinate way off the road. The location was between a cemetery (the company suspected that the insurgents often used the cemetery to store caches of weapons) and a farm. In the middle was a narrow strip of land between the two. Because of the laws of war, the Army's rules of engagement, and basic human decency, Ross and his patrol were not going to drive over the

cemetery or destroy the farm. They were forced to drive down this narrow piece of land, something the insurgents clearly knew. Ross recalled
that "we were traveling very slow, less than five miles an hour, as slow
as a Bradley fighting vehicle can go. We were creeping along; there was
a lot of 'moon dust.'"

This was the soldier's apt description for the very fine Iraqi sand
that penetrated everywhere. When soldiers walked through it, their
boots sank into it as if they were in quicksand. The area was coming
out of summer, so not all the of moon dust had blown away yet. The
Bradleys' tires drove the fine sand up in the air like the bow wave of a
ship in heavy seas. As Ross recalled,

> I remember yelling "stop" to the driver, Specialist Kurt Kammerick,
> just before the explosion occurred because I felt in my gut that
> something wasn't right. I could tell we were getting canalized and
> needed to stop. As we got closer, I didn't see anything like what we
> were looking for. There was no indication of anything. There was
> nothing, no people out; the farm was empty. Everything about the
> situation gave me a bad vibe. So, I was going to cancel this then and
> there, but it was too late. After I said "stop" the vehicle slowed until
> the explosion occurred.

This was an underbelly attack. The Bradley was a flat-bottom vehicle,
very vulnerable to an attack such as this because its chassis hung low
to the ground. Even though an extra layer of armor had been welded
on to the bottom of the Bradley, it could not limit the damage caused
by an IED of this enormous size. The insurgents intentionally offset
the pressure plate in order to hit the center and back of the Bradley
and kill or wound the most U.S. soldiers. Later Army analysis estimated
that the IED was in a fifty-five-gallon drum buried in the ground and
filled with 560 pounds of homemade explosive. Sitting in the cabin of
the rear Bradley, Private Tony Martinez felt the familiar concussion of
an IED, but he knew that this was much louder than anything he had
ever heard before. This was a massive explosion.

The lead Bradley shot up in the air; pieces of metal flew in all directions. Most of the armor tiles were blown off. The explosion was so intense that it blew off the Bradley's 1,500-pound hatch cover, sending it flying more than a hundred feet past the front of Sergeant Zamarippa's Humvee. He thought that no one in the lead Bradley could possibly have survived and so immediately set up a perimeter to help get the rest of the patrol back into the fight. Zamarippa looked and saw Ross through the smoke and dirt helping his driver and his gunner out of the stricken Bradley. Zamarippa described it as the most amazing thing he had ever seen.

Inside the Bradley immediately after the explosion, there was horrifying mayhem. In the back, the three soldiers were killed instantly: Sgt. Dan McCall, age twenty-four, from Pace, Florida; Pvt. Rush Jenkins, age twenty-two, from Clarksville, Tennessee; and Pvt. Cody Carver, age nineteen, from Haskell, Oklahoma, who had been in Iraq for just over thirty days. In the front cabin, there was a thick cloud of dust. Ross couldn't see anything and wasn't sure if he was alive or dead. He didn't know what was going on.

My head had gone through the FBCB2 computer screen that sat in front of me. I looked over and began feeling for my gunner, Sgt. Victor Larunde. I called out his name. He wasn't responding at first, but I shook him and he answered me. He was bleeding all over his face. It was chaotic. I called for the driver but couldn't hear him, and no one was responding from the cabin behind the turret. I poked my head out; couldn't see anything outside the Bradley because of the cloud of smoke, dust, and dirt. Fuel was dumping everywhere.

Ross exited the Bradley by pulling himself out of the top turret hatch. He jumped to the ground from the top of the hatch. Normally, soldiers don't do this because of the risk of breaking a leg, but Ross's adrenalin was pumping. Sgt. Victor Larunde, his gunner, was severely injured. The impact of the huge IED explosion had blown his face into the eye-piece periscope that sat in front of him. It had crushed his nose and eye, leaving him unable to see anything. Ross stood there controlling the situation. He helped Larunde understand what was happening and helped him get out of the wrecked Bradley. As Ross remembers, "I helped my gunner, Victor Larunde, out of the hatch and helped him down. My driver, Kurt Kamerik, had been thrown from the vehicle; his hatch had been blown off, and he was unconscious on the front of the Bradley. But he came to. We got down off the Bradley, and we were heading back. Sergeant Zamarippa had got out of his vehicle and was running towards us through the cloud of dirt and smoke. He thought all of us were dead. No one in the platoon expected anyone to survive the explosion that had occurred."

Thankfully, Kamerik and Larunde both lived, although they both had to be medically discharged from the Army because of the seriousness of their injuries. They came under intense enemy fire that persisted for hours as Ross and his platoon attempted to secure their location and get support assets in to remove the hulk of the Bradley. Later, both Kamerik and Larunde said that Ross had saved their lives.

Meanwhile, Sgt. Tom Cotterell, in command of the second Bradley at the back of the patrol, got out and moved quickly to restore radio communication, the patrol's lifeline. He ran to Ross's disabled Bradley to get the power amp, took it to his undamaged Bradley, and restored communications with Captain Clark and the platoon sergeant, Pete Black.

Ross was suffering from a concussion, a traumatic brain injury, a serious neck injury, shock, and multiple lower-extremity injuries. Despite his injuries, he took charge of the situation, consolidating his patrol back at Sergeant Zamarippa's Humvee. As they did so, they began receiving incoming machine-gun and small-arms fire from the farm house, the same farm Ross steered away from to avoid damaging its crops. He quickly realized that he couldn't get his rifle or his Kevlar helmet. His Kevlar had been destroyed in the explosion; his rifle was in the back of the Bradley with the bodies of his three dead soldiers. One of their arms was hanging out of the back of the Bradley where the troop hatch used to be. Inside, the bodies of the three young soldiers were unrecognizable. The explosion had pushed the floor of the Bradley up to the roof, and the metal was badly twisted. This damage and the position of the Bradley, perched precariously on the upper edge of the crater, made it impossible for anyone to get into the back of the Bradley to recover weapons or check on the three soldiers.

While Sergeant Zamarippa organized the defense, Ross reprogrammed Sergeant Zamarippa's radio to report the attack and the casualties to higher command. Ross told Captain Clark, his company CO, that he needed additional assets, including close combat aviation, to eliminate the current threat. Ross also checked on the well-being of each soldier in the patrol. At this point his medic, Doc Biddle, gave him an update on the two seriously wounded soldiers and the three who were apparently dead. Once the Apaches were on station, Ross coordinated their attacks against the insurgents.

Despite his wounds, Ross did not think he was really injured until Biddle stopped him and kept trying to get his attention. Biddle finally succeeded by yelling at Ross using his first name, something he would never normally do: "Ross, I am talking to you." Biddle knew he was out of line but was doing it to get Ross's undivided attention. Biddle said, "I need to look at you and evaluate you." Ross replied, "Okay." Biddle checked Ross over and asked him some questions, and as enemy fire came in, Ross answered as best he could. The medic concluded that Ross needed to go back and get a full medical workup. Ross replied, "There is no way I can do that right now. Let's continue to fight this fight and we'll worry about that later."

Ross was right. The patrol was under heavy sustained enemy machine-gun and small-arms fire. They could not leave the Bradley, nor could they get the bodies out of the back of it. There was nothing they could do but defend their position and coordinate with the Apaches. Guided to their target by Ross, the Apaches were very effective, but the firefight became more protracted because the insurgents kept gathering in more fighters now that they had U.S. troops pinned down.

Soon the firefight developed its own rhythm, with intense fire followed by lulls. The insurgents would renew their attack on Ross and his remaining soldiers with increased intensity. Tony Martinez, one of his soldiers, observed how calm and in control Ross was despite the chaos and the violence surrounding him. Tony described it as "a rare attribute that not a lot of men have."

Reinforcements began to arrive quickly at Ross's besieged position. First on the scene was a section from a sister platoon. They helped Ross and Sergeant Zamarippa build a defensive perimeter. Next to arrive was Sergeant Black and the rest of Ross's platoon accompanied by Captain Clark. They had been deeply shocked by the scale of the ambush. Black thought, "This can't be happening. These are my kids." To this day he remains affected by what he saw that morning. A tough combat veteran, he could not believe the horror now so visible in the back of the Bradley, which looked like a crushed soda can. Black covered the bodies of the three deceased soldiers with a tarpaulin as soon as he could.

Captain Clark had mobilized every asset he had to help Ross's besieged patrol, redeploying a platoon conducting operations in downtown Salman Pak as well as a platoon resting at COP Cahill. He directed them to come together and then come at the ambush from the far side. At this point the firefight with the insurgents had been going on for nearly three hours, and one of the reinforcements had been shot. Despite suffering from concussion, a brain injury, a neck injury, and multiple lower-extremity wounds, Ross led the defense of his exposed position with skill and fierce determination.

When Captain Clark arrived, Ross gave him a situation report. Clark told Ross that he agreed with Biddle that Ross needed a full medical workup and should be evacuated. To this day, Ross regrets not arguing with him. He wishes he had stayed and continued to fight, but he obeyed Clark's orders.

Ross helped Kammerik and Larunde get in the back of another Bradley, and they began their journey to a safe site where a medevac helicopter could land. To avoid another IED attack, Ross told the new driver where his patrol had come in off the road and directed him to take the same route back to the main road. The Bradley commander and driver acknowledged Ross's direction, but then somebody—possibly the

Bradley commander—overruled him. The driver took a shortcut. Ross was in the back with Kammerik and Larunde. Ross found a spare Kevlar helmet, which he immediately put on and buttoned up. He looked over at his two soldiers who were badly wounded. They were lying down and feeling miserable, dehydrated and in a lot of pain and struggling to deal with concussions. Ross told them to put their Kevlar helmets on. They were badly wounded and said, "Sir, right now we just need to get back." Using some "pretty colorful language," Ross ordered them to put their Kevlar helmets on. Right as they finished snapping their Kevlars on, their Bradley hit an IED. The three of them were thrown violently around the narrow confines of the cabin like "beans in a can." Had Kammerik and Larunde not had their helmets on, they would have suffered further traumatic brain injury and may have been killed.

The three of them were lucky, very lucky. The second IED was roughly the same size as the first, and once again the insurgents detonated it beneath the rear troop-carrying cabin to cause maximum U.S. casualties. But this time it detonated at an angle. The Bradley was decanted, driving on an incline, and so much of the force of the explosion went off to one side. It collapsed a portion of the Bradley's interior but did not take the whole floor and push it up against the ceiling as had happened in the attack on Ross's Bradley. The Bradley's commander and driver were badly shaken up, but the driver's hatch was not blown open, so unlike Kammerik, the driver of this Bradley was not thrown from the vehicle. The IED still destroyed the Bradley, but they all survived. Kammerik and Larunde say that Ross saved their lives again. With his usual wry humor, Ross says that "they saved their lives because I was going to kill 'em if they didn't put their damn helmets on."

When they got out of the crippled Bradley, there was more insurgent machine-gun and small-arms fire. Ross helped Kammerik and Larunde down on the side of a berm they had been driving on. It was as much cover as he could give them. As the enemy gunfire continued, Ross looked for enemy positions, especially observers on roofs. At this point he could see that they were being watched as other insurgents moved in on their position from multiple directions. He was trying to identify where exactly these hostiles were coming from so he could guide the fire from the gunners of the other Bradleys onto the enemy.

The firefight continued intermittently before an engineer team led by Sgt. Jeremiah Gann cleared a path for Captain Clark, Sergeant Black, and others to return safely to COP Cahill. This path also enabled a third Bradley to come in to take Ross and his two badly wounded soldiers to a safe site where a medevac helicopter could land. The medevac helicopters took them to the U.S. military hospital in Balad, fifty miles north of Baghdad.

As Ross, Kammerik, and Larunde were being evacuated to Balad, the firefight at the ambush site continued, lasting for another three hours. One other U.S. soldier was slightly wounded. Repeated Apache attacks coupled with intense, accurate machine-gun and rifle fire from the U.S. troops forced the Iraqi insurgents to give up and withdraw. When the U.S. troops left, they took everything with them; nothing was left behind. The two severely damaged Bradleys were loaded onto flatbed trucks. Company mortar platoon leader Capt. Chris Pearson personally supervised the recovery of the dead solders' bodies.

Balad Air Force Theater Hospital was a major trauma center, the biggest U.S. military hospital in Iraq staffed by dozens of highly skilled, exceptionally dedicated surgeons and nurses—the best of the best—who saved hundreds of American and Iraqi lives. The Balad hospital was not a brick-and-mortar building like its civilian counterparts in the United States. Rather, with the exception of one low-rise wood frame building, it was a series of interlocked tents covering 35,000 square feet; surgeons carried out their lifesaving procedures in sanitized shipping containers.

When the helicopters landed on Balad's helipad, medical orderlies rushed Ross, Kammerik, and Larunde into the intensive care unit for immediate assessment. The medical staff cut everyone's bloodstained uniforms off and threw them away. They gave Ross and his two soldiers multiple X-rays and scans. During that first day, Ross was able to call April to let her know "I'm fine, I'm doing okay, I'm alive, I've got all my digits." That call set in motion a remarkable chain of events that speaks volumes about the selfless moral heroism of the Pixler family.

At 9 p.m. on October 30, 2008, Ross's father, Reid Pixler, had just returned to his office at the U.S. Army base in Mosul, Iraq, after attending mass celebrated by a visiting Army Catholic chaplain. Reid, assistant U.S. attorney for Arizona, was serving as a Department of Justice resident legal adviser to the Iraqi judicial system and Rule of Law Team leader for the State Department's Provincial Reconstruction Team. based in Mosul. His mission there had been twofold. The first was to create a secure judicial compound where terrorism trials could be safely conducted. The second was to establish a rotating system whereby Iraqi judges could be brought into Mosul from around the country to hear terrorism cases.

Just after 9 p.m., Reid's office telephone rang: it was his wife, Larissa, calling from Tucson, Arizona, over a State Department satellite link. She told him that Ross had been badly wounded in combat and wanted to see his father. Larissa had learned the bad news from Ross's wife, April. She had told Larissa that Ross was not allowed to describe the nature of his injuries, but the good news was that he was talking. He was asking for his father; could Reid get to the Balad military hospital to see him? On the call with her husband, Larissa said, "You need to get to Balad." Reid, carrying many onerous responsibilities, replied, "I can't just go to Balad." Larissa screamed at him, "Get to Balad!"

After the call with Larissa, Reid called his Provincial Reconstruction Team's doctor to get confirmation that Ross was in the Balad military hospital. Through his medical network, the doctor was able to confirm that Ross was in fact there and he was injured but was unable to get any estimate of the extent of his injuries. Once Reid had confirmation, he wrote his own travel orders, took his "go bag," a military duffel bag full of basic essentials, and headed to the Mosul military airfield to see what flights were available to Baghdad or Balad, nearly 250 miles away. With sectarian murders, suicide bombings, and attacks on U.S. personnel as a daily occurrence, driving was not an option.

At the check-in desk, Reid explained who he was and that he was trying to get to Balad to visit his wounded son. The Air Force enlisted man at the desk told him that there were three potential flights that night. The first was a special forces flight of two Chinooks, but their mission was top secret, and besides they often landed at the far end of the runway without notifying the control tower. The second option was a C-130 transport plane that was due to fly on to Balad, but that wasn't confirmed. The final option was another C-130 that was dropping Army personnel off in Mosul before flying directly to Balad. The threat of ground-to-air missiles and other antiaircraft fire was so prevalent that C-130s did not stop and turn their engines off. Instead, when they landed they did a speedy off-loading and boarding with the engines running.

Reid now began to try to get a seat on this C-130, which had recently landed. There wasn't much time. Reid knew the crew chief on this C-130 well from the many trips he had to make from the Mosul air base. In Reid's words, he was a "wonderful man." Reid explained why he was trying to get to Balad urgently. The C-130 crew chief replied, "Reid, there are twelve men with confirmed seats on this flight, and there are twelve men going to get off. Since you are the equivalent of a colonel you can bump one of them off, but you will be interfering with some-body's orders." Reid said, "I can't do that, but I would like to be first in line for a space available after that." The crew chief went over to the nearly thirty military personnel waiting for a flight. The first twelve had confirmed seats, and the remainder were flying standby. When the crew chief explained to them that Reid was trying to fly to Balad to visit his wounded son, every one of the eighteen stand-by military passengers picked up their gear and took a step backward. Reid was now number one for standby, but his night's adventure had just begun. The crew chief ushered his twelve confirmed passengers and Reid on board the C-130, which took off from the Mosul airfield at about 11 p.m.

Immediately after takeoff when the aircraft began to climb to its cruising height, insurgents "painted" the C-130 with a laser designator, an electronic targeting device for a ground-to-air missile. The plane's internal alarms began to ping. The pilot took vigorous evasive action, and the plane began to pitch violently. Reid looked out the window and thought that the starboard engine was on fire. He saw red, orange, and yellow streamers coming out of the back of the engine. For a moment,

Reid thought he was going to die. But he quickly realized that the pilot was deploying countermeasures flak and flares to distract the trajectory of the surface-to-air missile. The C-130 landed safely at the Balad air base just before 2 a.m. For Reid "it was a roller coaster ride, the wildest ride I ever had on a C-130."

Reid's Provincial Reconstruction Team's doctor had called ahead to the hospital to let the medical staff know he would be arriving that night. As a result, a military police officer with a jeep was waiting for Reid when he arrived, and he was driven directly to the military hospital.

When Reid came through the main entrance escorted by the military police officer, straight ahead was a general-purpose room with a television and a few chairs. The first thing Reid saw was Ross sitting in a wheelchair, and that "freaked him right out." He thought, "Oh my God, how do they have him in a wheelchair already?" In fact, the explanation was benign. The room only had a few chairs and sofas, and all of the free seats were occupied. Ross wanted to watch television, and the wheelchair was the only seat available.

Reid remembers what he saw after that initial shock. "When I turned left, I faced a small nurses' station. As I recall, a hallway ran to my left to an area that I did not explore. Straight ahead were two rows of hospital beds occupied by soldiers, one row on the left and the second on the right. There was some form of fabric curtain that somewhat separated the beds from each other, at least partially. There was enough room between Ross's bed and the curtain for a small rollaway bed the nurses set up for me."

Ross was surprised and overjoyed to see his father. Later, he remembered that his father "found a way to me, that's for sure. That was really special." Father and son wrapped each other in a prolonged bear hug. Ross told Reid what had happened. He did not think he was seriously injured, but he was aware that the doctors had concerns about his neck. After a long conversation, Ross and his father fell asleep in adjacent beds at about 3 a.m. The nurses were excited to see Reid because it was the first time they or any of the soldiers in their care had "a dad visit." For the next two days, Reid became "dad" to thirty wounded soldiers.

In the middle of that first night together, there was a horrible racket inside Ross's ward. A wounded Iraqi soldier in the bed across from Ross began screaming in Arabic. It woke Reid first. He didn't know where he was or what was going on. Groggy and sleep-deprived, the one thing Reid knew was that he shouldn't be hearing Arabic. Instinctively he reached for his gun, but the first thing he saw was Ross getting out of bed and walking gingerly toward a young Iraqi soldier who had taken the full force of a hand grenade explosion to the face. In the explosion, he had lost his right arm and was now blind. U.S. Army surgeons had dug the shrapnel out of his torso and his face, but this young soldier was now maimed for life. He woke suddenly in the night and began screaming because he didn't know where he was and had become tangled in his IV. He was terrified. Despite his own injuries, Ross was stroking the young Iraqi soldier's head, speaking softly to him in Arabic. When the nurses arrived, Ross explained what had happened and told them that the young wounded Iraqi was very thirsty and needed water. Ross stayed with him until the young Iraqi calmed down and went back to sleep.

During his time in Balad, Ross felt very guilty lying there in the hospital. He knew he had a concussion, a brain injury, and some short-term memory loss. But as he lay in his hospital bed, he kept thinking, "I have no excuse for being here. I have no life-threatening injuries. I should be back out there." Ross wanted to be back with his soldiers, who were still fighting the enemy insurgents despite the loss of three comrades. They needed comfort, support, and leadership, all of which Ross could have provided so effectively had he not been hospitalized.

Ross's doctors remained concerned about his condition, however, and wanted to fly him to the U.S. Army Medical Center in Landstuhl, Germany, for further evaluation, including an MRI. Ross refused. He remembers thinking, "I feel this guilty now. Imagine how I'll feel if I get to Germany and there's absolutely nothing wrong with me. I will have left the country for good and left my soldiers back there to continue fighting when the violence has escalated beyond anything we had seen. Somehow, if I live through this. I need to get back to my unit." So, Ross told Kammerik and Larunde that he wanted them to go to Germany but that he would not be going with them. They wished each other well.

Ross's decision to remain in Iraq reflected his deeply felt moral values and his unswerving, selfless commitment to his soldiers. But it came with a price. The Balad doctors' concern was well founded: in addition to a serious concussion and a mild brain injury, Ross also had a severe neck injury. Without an MRI, the Balad doctors were unable to determine precisely from X-rays alone how serious that injury was. Ross himself would only discover the seriousness of his neck injury a year later, in 2009, when he found that he was unable to use his right arm or move his neck. He had had persistent pain but had just lived with it. After fierce lobbying of the Army bureaucracy by his wife, April, Ross finally received the MRI and the surgery he needed. The MRI showed three bulging discs and two herniated discs in his neck. In June 2009, at Saint Francis Hospital in Columbus, Georgia, he was operated on by Dr. Bill Adams, one of the leading neurosurgeons in the country. Dr. Adams installed a metal plate in Ross's neck, which successfully resolved the disc problems. Dr. Adams later told Ross that had he received any further trauma, he could have likely become a quadriplegic.

Determined to get back to his unit, Ross asked to be released from the hospital. The doctors disagreed. They insisted that he needed to go to Landstuhl and receive more tests. So, Ross left the ICU and found a liaison officer who provided him with a fresh clean uniform, although it did not have any patches, rank, or insignia normally worn by soldiers. Ross left the hospital and walked to the military airfield. This was the largest U.S. military air base in Iraq, the heart of U.S. Air Force operations in the country as well as the center of U.S. Army logistical operations, the enormous Anaconda Logistical Support Area.

Ross reported to the Air Force check-in desk for military personnel needing transport to their unit. The soldier behind the desk told him they had two roll calls per day: ten in the morning and ten at night. So, Ross should return at one of those times to get on a flight. Because he had no identification and no orders, Ross did not want to draw attention to himself by walking around the base and potentially getting pulled back to the hospital. He waited at the airport injured, in pain, and hungry. Technically, Ross was now AWOL, loose inside the Balad Air Force base with no common access card, no identification card, no orders, no uniform insignia, no weapon, no body armor, no Kevlar helmet, and no means of communicating with anyone.

At 10 p.m. Ross returned to the check-in desk, but the soldier on duty told all of the soldiers waiting that no more flights were going out that night and that they should all return the next day at 10 a.m. As everyone turned and started walking away, the Air Force desk officer tapped Ross on the shoulder and said, "We usually try to help out you guys. Where are you trying to go?" The desk officer incorrectly assumed that Ross was with one of the many special operations units operating in Iraq. He pulled Ross into a back room to where a map was hanging on the wall. Ross told him he was trying to get to COP Cahill. The soldier at the check-in desk replied that he didn't know where that was, so he asked, "Where is that near? Where is your center FOB?" Ross replied, "FOB Hammer." The Air Force sergeant said he didn't know where that was either. It was a new FOB, and he was not familiar with it. Ross asked, "Do you know where the Butler Range Complex is?" The sergeant said, "Yes, of course, we know where that is. It's the old Republican Guard Tank Range." Ross replied, "That is where it is, at the southern tip." The soldier said, "Okay, we can get you there." Ross accepted because he knew that he could get a helicopter flight from FOB Hammer to COP Cahill. "Where's your ID, your orders, your weapon, and the rest of your gear?" Ross replied, "I don't have any of them. I need to get back to FOB Hammer to get them." The desk officer consulted the crew chief and the pilots of a Black Hawk helicopter waiting to take off. They said they could not allow Ross on the flight because it was against all sorts of regulations to carry a passenger without identification, orders, a weapon, and the rest of his gear. But since they wanted to help a special operation soldier who might be on a clandestine mission, the pilots told the crew chief "hold him back in the shadows over there. Hide him. And if it's safe we'll waggle the landing lights on the back of the bird. Then you bring him out, and we'll try to get him out of here." That is what happened.

Ross boarded an Army Black Hawk on one of the nightly ring route flights, a predesignated flight path for helicopter pilots flying at night that was designed to prevent midair collisions with other U.S. military aircraft. These nightly transport flights usually involved bringing soldiers to and from different combat outposts. Because FOB Hammer was not yet on the nightly ring route flight path, the Black Hawk pilots had to add an extra stop for Ross at the end of their legs. This meant that he had to fly around the whole ring—a five-hour journey—all over Iraq.

When the Black Hawk landed to refuel, a colonel got off and saw Ross sitting in the back of the Black Hawk cabin and screamed at him "What are you doing? You don't have any of your gear. Where is your helmet and weapon?" When the colonel paused to take a breath, Ross said, "Sir, I would be lying to you if I told you that this was the first time I've flown without any gear or equipment." Ross was relieved that the answer was enough to satisfy the colonel's curiosity. He turned away mumbling about not wanting to know anything about what Ross was doing.

Around 3 a.m., the Black Hawk dropped Ross off at the Butler Range Complex, not at FOB Hammer. Ross didn't realize this until the Black Hawk flew away, his eyes adjusted, and he began to get his bearings. Once he did, he realized that he was now behind enemy lines.

Once the helicopter took off, Ross looked around but could see nothing in the inky blackness. He wasn't on any kind of concrete. In fact, he was in the middle of the desert, and the lights of FOB Hammer were nowhere to be seen. Ross realized that he was in enemy territory probably somewhere in the vast Butler Range Complex. Over the previous months, insurgents had attacked FOB Hammer from the northern portion of Butler Range Complex, probably because they knew that the American soldiers were not willing to enter the impact area of an old tank range due to the threat of unexploded ordnance. On one such attack, the insurgents had launched twenty-six Katyusha rockets at the base from the general area where Ross was now standing. He was well aware of this attack, since one of those rockets had wounded Private Phillips, one of Ross's soldiers, a few weeks prior while he was waiting for a flight to get back to COP Cahill. Ross knew he was standing in a minefield of unexploded ordnance. Worse still, he had no GPS, no compass, no weapon, no communications, no Kevlar helmet, and no body armor. And he was still limping. Ross stopped to think. In his mind he began to reconstruct the geography around FOB Hammer. The Butler Range Complex was diamond-shaped. FOB Hammer was in the southern part of the complex. He decided to walk south. It was a relatively clear night, so he searched for and found the North Star and walked away from it.

Because Ross was in enemy territory, he walked about a dozen steps and then stopped. He knelt down to listen carefully to check if anyone was around. He thought that if he found an enemy insurgent, he could kill him with his hands and take his weapons. But if there were more insurgents, he would have to hide. Ross could hear nothing, so he resumed walking. Injured, exhausted, and hungry, he walked three miles across the desert minefield in the pitch darkness, stopping every few minutes to listen, smell, and look for signs of enemy activity.

The first structure Ross encountered was the Iraqi military base to the north of FOB Hammer. In the towers, the Iraqi Army guards saw this man limp into the area, lit by their perimeter arc lights. They were understandably suspicious, pointed their rifles at Ross, and began shouting at him in Arabic. Ross put his hands up and shouted "*mutar-jim*," Arabic for "interpreter." It was the only Arabic word he could think of at that moment, but it worked. The Iraqi soldiers did not fire. Instead, they got an interpreter, who walked out to Ross, who was still standing with his hands up in the desert. He searched Ross and started asking him questions about who he was and what had happened. Satisfied that he was an American, he brought Ross inside the Iraqi base and let him call FOB Hammer. The Iraqi interpreter told Ross that he was lucky he had not been shot by the guards. They often tended to shoot first and ask questions later. Within thirty minutes, two American Humvees came to collect Ross. Once back at FOB Hammer, he got a helicopter flight to COP Cahill.

At COP Cahill, Sergeant Black was surprised to learn that there was an incoming helicopter. Neither he nor Captain Clark were expecting a supply run. When they learned that Ross was on board, they were stunned. After Ross landed, Black looked at him and said, "What are you doin'? You need to be checked out, man. You need some time to decompress, man." Ross replied, "I need to get back to work." Black looked him in the eye and said, "I don't care what you want. You need to get back on that helicopter and get your butt to the hospital." Clark overheard the conversation and came over and said to Ross, "What are you doin' here?" Ross replied, "Get me back to my men. I am not bleeding. I still have all my fingers and toes. I have a brain. I can move under my own power." As Ross finished speaking, Black thought, "I can't fault the guy. This dude really cares."

Captain Clark welcomed Ross but made it clear that he would not allow him to resume his normal combat duties. His soldiers were not surprised by Ross's selflessness but could also tell that he had not recovered from his injuries. Sergeant Black suggested that Ross go home for a bit to recover. So, Capt. Clark assigned Ross to some routine on-post duties until he could catch the next rest and recuperation (R&R) flight out. Because of his hospital stay, Ross had missed his scheduled R&R date anyway, so Clark's decision provided the perfect solution.

It was also the perfect decision in another way. Ross was still physically and psychologically wounded from the insurgent ambush. His physical wounds were healing, but psychologically he did not have closure. He had missed the funerals of his three soldiers killed in the IED explosion and had not been available to comfort and support the rest of the platoon as he wanted. He needed time to heal.

In early November 2007 Ross returned to his home base at Fort Benning, Georgia. For Ross it was a warm, loving reunion with April and his baby daughter, Dakota, who was celebrating her first birthday. It was great for April, who could now see and touch him. There was also an opportunity to see and visit with more members of his family at a long-planned extended family get-together hosted by his uncle, Leon Tasheiko, at a rented house in Fair Hope, a charming old southern town on the Alabama coast. The occasion was Leon's wife's birthday too. Nobody had expected Ross to be there, so it was a wonderful surprise for his aunt, uncle, and other relatives. These were extended family he didn't see very often. Indeed, it was the first time he had seen many of them in years. It was a pleasant and enjoyable visit, a chance to rest in the warm Gulf Coast sun.

But for Ross it was also very difficult. Physically he was in Alabama. Mentally he was still in Iraq. He had just suffered a serious trauma: hit by two massive IED explosions in one day, causing multiple injuries. He had also suffered the deaths of three of his soldiers and the serious wounding of two more. Like every member of his close-knit platoon, Ross was devastated by these losses. He also felt guilt-ridden sitting there in the Alabama sunshine while his soldiers continued the fight in Iraq. The problem was that his relatives had no framework, no basis for understanding what had just happened in Iraq, and Ross was not yet ready to talk about it.

Immediately after his R&R leave, Ross returned to his platoon and resumed his normal duties. Sergeant Black had kept everyone in the platoon focused on their mission and prevented a decline in morale. What helped was killing some of the insurgents who had planned and executed the ambush and capturing some of the others involved. Intelligence provided by a local sheikh led to a source who gave Hardrock Company the information they needed to identify those responsible for the five casualties in Ross's platoon.

After his return, Ross recalls seeing the prisoners before they were moved to the detention holding facility. He was stunned by how young they were—in their early twenties—and had expected they would be older. Ross remembers feeling "great hate" toward the prisoners. He wanted to see them dead, but he also wanted to ensure that none of his soldiers did anything that would later compromise their career. These insurgents were not worth anyone's career.

For Ross, helping the platoon heal was his highest priority. He recalled that "because we were so close, it [the IED ambush] hit us so hard. There wasn't a person in the platoon who wasn't devastated by it. But we were there for each other; we were inclusive. There was no in-clique or out-clique. We were all together. We always did everything together. So, we were counseling each other in the gym. When we had smoke breaks, Staff Sergeant Cottrell would always play his guitar. We were able to deal with it in our own ways." It helped that by this time the platoon had, in Tony Martinez's words, "gained a callus. We understood what it was to lose a brother, to suffer through injury and keep driving on. It was something we had already accepted. We knew death was a possibility."

Captain Clark agreed with Ross that the platoon needed more help in dealing with their grief and requested that the brigade psychologist, Capt. Angela Mobbs, come to COP Cahill. When she arrived, she was unable to get members of the platoon to talk to her. The soldiers were reluctant to speak to her because of the stigma attached to admitting that they needed mental health counseling. So, Ross issued an order: everyone including himself had to talk to Mobbs. He went first and last. His action encouraged all of his soldiers to open up and, because of his stature within the platoon, gave legitimacy to seeking mental health counseling. In his concluding meeting with the Brigade psychologist, Ross asked for and received the advice he needed to ensure the health and safety of everyone in the platoon. At his request, she returned a couple of times to give Ross some further useful tips on how to approach individual soldiers, and she put some of them on medication. After further evaluation, she and Ross concluded that one soldier in particular needed additional support. They sent him for further treatment, including group therapy. After he returned, he pleaded with Ross never to send him to group therapy ever again.

Within the platoon, Tony Martinez observed that Ross was still the same selfless leader as before and was still one of the guys. But after the ambush of October 30, "he cared that much more; he took the missions even more seriously because he knew that within the blink of an eye anything could happen."

In November 2007 as part of the Pentagon's wider policy, Ross's unit's Iraq deployment was extended so that he and his soldiers would now not return home to the United States until May 2008. "It wasn't too much of a shocker to find that out halfway through our deployment," Ross recalled. Going into Iraq in March 2007, Ross and his soldiers

suspected that their deployment would be extended from twelve to fifteen months, and they managed their expectations accordingly. There
was not a single member of Ross's platoon who did not think that the
additional three months "felt like forever." But the simultaneous wars
in Iraq and Afghanistan had stretched the U.S. Army to near breaking
point. They were short of troops, and extending deployment tours was
one of the few ways open to the Pentagon to bridge the gap between
what they had available and what was required.

The platoon now had seven months left. Those months would
prove to be very violent, but growing combat fatigue in the platoon was
offset by the fact that they were winning every firefight. In fact, within
their area of operations they were stopping the insurgency in its tracks,
killing or capturing most of the enemy combatants.

Between November 2007 and December 2007, insurgent violence
intensified. The period December 16–21, 2007, was especially violent,
with intensive firefights every day. The fighting was so intense that on
several occasions, Ross's platoon had to return to COP Cahill to get
more ammunition and then get back in the fight. During this period,
they mounted a series of clearing and shaping operations in multiple
neighborhoods. They pushed south from Al Bawi through Al Zelig and
Dera and eventually Kanasa.

On December 21, Ross's unit finally broke the back of the insurgency in their area of operations because they had done so many operations and had been so successful in hitting the right targets that the
enemy was collapsing. On December 22, the day after the last firefight,
Ross and his soldiers left Camp Cahill expecting another day of intense
combat. They were "shocked and amazed" to see the streets of Salman
Pak lined with people clapping, cheering, and waving at the U.S. troops.
Husbands held hands with their wives as they and other women lifted
up their babies. This was something Ross and his soldiers had never
seen throughout their entire deployment. "You could tell that everyone
knew it was over," Ross said. "The civilians in Salman Pak were ecstatic:
their enemy had left." In this area of operations, ably led American
soldiers had defeated the insurgents, and everyone in the community

knew it. Ross and his soldiers knew it from the civilians, not from any U.S. or coalition source. As he later told me, "Their intel was better than ours. We weren't even sure it was over yet."

Years later Ross remembered that victory fondly. "It did really seem a lot like a liberation. And what was special about it was that it got us to realize that the people didn't support them. The vast majority of people there were victims, not supporters of their presence. It was really special to liberate them. We got to feel a little hint of what U.S. and allied forces experienced in Western Europe in World War II."

Ross does not recall any contact with the enemy after December 21, 2007, until February 5, 2008. It was eerily quiet; after the intense fighting of the previous weeks, it was so strange to go for such a long period without a single hostile shot.

On February 5, Ross led his platoon on a clearance mission deep into the heart of the enemy's territory in Dera. They went all the way down into what had been the epicenter of enemy resistance; the shadow government of Iraq's headquarters. Insurgents fired an 81-millimeter mortar at them. It missed and ended up in the Tigris. The second mortar round they fired blew up on themselves. That was the last hostile round fired at Ross's platoon during their deployment. As he later recalled, "It was so strange that we had gone from such violence all the way up and into December. Then the pause, then February 5, and then nothing." They had won a small but important victory.

During the surge, the vast majority of U.S. and coalition forces were fighting a counterinsurgency war. Ross and his soldiers were not. He believes that historians and journalists who have written about the surge have not focused enough on the fact that within the counterinsurgency, there were pockets of conventional-style war. He has a point. In their area of operations, Ross and his soldiers were fighting a classic high-intensity conflict. They knew where the boundary of the enemy's area of operations was. They knew that when they crossed that line there would be fighting. Sometimes the insurgents crossed their boundary and attacked Ross and his soldiers. It was textbook high-intensity conflict.

Ross and his soldiers prevailed in greater Salman Pak for several reasons, some of them obvious, including superior technology, training, and discipline, but there was more. Some military historians have correctly suggested that within an infantry unit, cynicism and fear can be contagious and can destroy a unit's combat effectiveness. In Ross's case it was the opposite. What you can see here is, as he put it, a "contagiousness of bravery" with soldiers supporting each other, willingly going into an enemy buzz saw day after day, "getting up every morning throwing their uniform on saying let's go get some more." Ross and many of his soldiers anticipated that they would die at some point during the intense fighting. Every day they got up, they believed that it could be their last. Everyone was scared, but as Ross remembered, "What made soldiers even more scared was their brothers going out there and they not being a part of it. There was not a single person willing to raise his hand to say 'not me, I don't want to go.'" To be sure, this was in part an expression of the brotherhood of battle. What Philip Caputo described in *A Rumor of War*, his classic account of Vietnam, applies equally to Ross and his platoon in Iraq. Caputo wrote that the war was "a crucible in which ... American soldiers were fused together by a common confrontation with death and a sharing of hardships, dangers, and fears."[1] But it was also a tribute to the brotherhood that Ross had created within the platoon. His selfless devotion to his soldiers helped build and sustain the bonds of affection, loyalty, and trust upon which that mutual love rested.

On February 8, 2008, Ross received the Silver Star in former Iraqi dictator Saddam Hussein's Al Faw Palace in Baghdad. The ceremony took place in the largest room in the palace and was lit by what appeared to be crystal chandeliers, but when you touched them they were merely plastic. Ross was accompanied by his father, Reid, and by Staff Sergeant Zamarripa, who also received the Silver Star. Vice chief of staff of the Army Lt. Gen. Dick Cody presented the medals.

The morning after the ceremony, Ross, Staff Sergeant Zamarripa, and Reid attended a breakfast given by General Cody for all medal recipients. Cody took a personal interest in each soldier and even asked Reid about Ross's older brother Ryan, an Apache pilot who was deployed in Afghanistan. Cody blinked hard and said to Reid, "Your son must be in 1st/285 Arizona National Guard. He is replacing my son in Afghanistan." Reid later recalled that "I was blown away." The breakfast took place in a small conference room near Cody's quarters. Reid recalled that "General Cody was remarkable. He engaged Ross and every other medal recipient personally, spoke knowledgeably about what they had done and what he had found especially impressive about their actions." Speaking without notes, it was clear to everyone present that Cody had carefully read each citation and that his compliments were rooted in a deep grasp of the citations. Ross noticed and appreciated how Cody, the second-in-command to the highest general in the Army, had gone out of his way to understand what he and Staff Sergeant Zamarippa had done.

After Ross got back from R&R, Maj. Gen. Rick Lynch, the division commander, flew to COP Cahill and asked to have lunch with him alone. Lynch wanted to know the whole story. He did appreciate and laugh at the fact that Ross had been mistaken for a special operator. Lynch said he was proud of Ross and wanted to reaffirm that he and his platoon were doing good work and that his efforts and sacrifices mattered.

22

BOBBY SICKLER

Iraq, 2007–2008

Because of his high class rank at Fort Rucker, Bobby Sickler was able to choose his first duty assignment. He was determined to go to war as soon as possible, so he chose to join the 4-6 Air Cavalry (4-6 Cav) based at Fort Lewis, Washington, because he had discovered that it was deploying next to combat.

The 4-6 Cav was a unique squadron, one of a kind in the U.S. Army. It was an independent cavalry squadron that had no brigade headquarters. Instead, it reported directly to the I Corps at Fort Lewis. The 4-6 Cav was a large squadron with just over nine hundred service members, had "an extremely robust maintenance package," Bobby said, and was equipped with a combination of Black Hawk and Kiowa helicopters. This independent squadron was able to operate in an open, independent environment and was free to develop new ways of fighting in which Kiowas and Black Hawks would work together. This was a new concept that the Army briefly explored but then rejected in favor of standardized units under brigade control.

Freed from the immediate constraints of a brigade headquarters, 4-6 Cav's commander, T. J. Jamison, was free to develop and explore his own vision of how air cavalry should fight a counterinsurgency war. For an innovative thinker like Jamison, it was a perfect fit. For an eager young pilot like Bobby, it was an ideal environment to learn to be the best Kiowa pilot he could be.

Bobby joined the squadron in late October 2006. Lieutenant Colonel Jamison was an inspirational leader who gave and inspired deep personal loyalty in his subordinates. Jamison, a tall man with dark brown eyes, graying hair worn in a crew cut, thin lips, a hawk-like nose, and strong protruding chin, was a native of Oklahoma. He began his military career as an enlisted man in his state's National Guard. He earned his Army commission from Marion Military Institute in Alabama, where he was also allowed to attend and graduate from the U.S. Army Ranger School. He switched to aviation and, after graduating from Fort Rucker, deployed to West Germany near the end of the Cold War. There, he flew helicopter patrols over the Fulda Gap. It was here that he began to conceive his concept of continuous aviation presence that he would later develop as commander of 4-6 Cav. Before that he spent several years at the Army's National Training Center in Fort Irwin, California, first as a tactics coach and later as an attack battalion and cavalry squadron trainer.

Lieutenant Colonel Jamison was a charismatic air cavalry officer. Bobby later described him as "the most fearless warrior I have ever met." On one occasion in November 2007 in Iraq, Jamison was shot through his flight helmet, made of a thin carbon fiber shell covering Styrofoam padding. He landed his badly damaged helicopter at the 4-6 Cav's base at Mosul airport, took off again in another Kiowa, and flew back into combat.

Lieutenant Colonel Jamison was a dynamic "let's do this" kind of officer, a confident air cavalryman with swagger and exceptional technical and tactical proficiency. What inspired Bobby and other members of the squadron was not only Jamison's confidence and fearlessness but also his deep personal loyalty to everyone in the squadron. As Bobby later recalled, "I have never seen anyone in the Army more loyal to the people who worked for him than he was. That bred an esprit de corps in the unit and a level of loyalty to him that I haven't seen replicated since."

The Alpha Troop commander, Capt. Mike Olson, was also an inspirational leader. Olson was in his early thirties, older than the average troop commander. Physically, he was a short man just a little over five foot three, but he was "pure energy." Olson, a devout Mormon, was also a calm, humble, likable man with blond hair, a cheerful personality, and "endearing mannerisms." As Bobby told me, "You instantly liked the guy and would do anything for him. Captain Olson exuded a quiet authority. He never said and never needed to say 'I'm in charge, and you are doing it this way because this is what I say.'" He was a highly effectively commander, instantly building trust with the people who worked for him. Olson had a major impact on Bobby's development as a junior officer.

Within the troop, there were a number of good warrant officers. In Bobby's view, the standout was the senior warrant officer three (CW3), Tom Boise. A former special forces NCO, Boise had gray hair worn in a crew cut, striking lightning-blue eyes, and a ruddy complexion. He exuded confidence and leadership and was a "pure warrior" who if someone said "maybe we shouldn't go out scouting because we might be shot at" would reply, "As you get shot at, you know where they're at." Boise had a big influence on the way Bobby understood combat.

CW3 Boise had a knack for adjusting tactics to the environment and creating innovative ways to accomplish the mission that were adopted by the rest of the squadron. For example, the dusty air sometimes clouded the Kiowas' sights. Boise developed a way for Kiowas to fly together as a team, which consistently increased the chances of their missiles hitting their targets. As Bobby recalled, Boise's "ability to adjust our tactics to the environment and the enemy had a profound impact on the way I understand war even today. He constantly treated combat as an interaction with the enemy rather than a set of sets that didn't allow for any input from the enemy or the environment."

At Fort Lewis, Bobby's first assignment in the squadron was to serve as the supply platoon commander for the maintenance troop. Lieutenant Colonel Jamison gave him this assignment because he believed it would help him develop knowledge he would later need as a troop or squadron commander. In late January 2007 Bobby transferred to Alpha Troop, a regular flying troop.

Lieutenant Colonel Jamison already recognized Bobby's potential and assigned him there because he wanted a cluster of outstanding lieutenants to support the new troop commander, a newly promoted and untested captain, Mike Olson. Olson had had to take over Alpha Troop under awkward circumstances. Jamison had relieved his predecessor for allowing the troop to drink alcohol during training exercises.

The squadron's deployment to Iraq was accelerated significantly, and they did not have time to go to the National Training Center for a full predeployment train-up as they normally would. Instead, the National Training Center sent a team to Fort Lewis to try to replicate the Fort Irwin experience. It was not ideal, but they did their best. Time was short, and Bobby only had the minimum allowed by army regulations to prepare to deploy to Iraq.

The first challenge was to integrate into the troop and the squadron before they deployed to Iraq. Integrating meant building trust with a group of pilots, most of whom were combat veterans and had already established those special bonds that develop between soldiers who have fought together. Within the 4-6 Cav, about two-thirds of the pilots were veterans of Iraq and Afghanistan; one-third were, like Bobby, new arrivals. Most of the squadron who had deployed had fought in Iraq as part of the initial 2003 invasion. The reminder of the veterans had served in Afghanistan with the 101st Airborne.

Lieutenant Colonel Jamison had created an outstanding squadron culture, and his officers, NCOs, and soldiers were, in Bobby's words, "incredible professionals." He later recalled that "they welcomed me in. They recognized that I was a young lieutenant who was still learning my job. I leaned heavily on those guys, because I knew what I didn't know." Bobby's approach to his new colleagues was "I am here to learn from you guys who have done this before."

His second challenge was to become as combat ready as possible before he deployed. To that end, he completed the additional training in the squadron that Fort Rucker required and flew with members of his troop as often as possible. Thanks to his own single-minded focus and the mentoring of his commanders and colleagues, Bobby made rapid progress. It was all according to regulations, but time was limited. As he later recalled, "I was hanging on for dear life."

Captain Olson devised some imaginative training exercises that were instrumental in Bobby's development. One example was when the troop went out to the Fort Lewis urban training center, a replica of an Iraqi town. Bobby's job was to play the role of the insurgent, "the bad guy on the ground" driving a van. As part of the exercise, he hid the van between buildings and hid himself next to windows in the buildings where he pretended to fire on the Kiowas. Bobby now had a clear idea of what insurgents on the ground could see as well as when and from where they could fire at U.S. helicopters. "It was an experience that helped me when I got into combat," Bobby noted later.

The troop and the squadron continued to give Bobby every opportunity to learn the combat skills necessary for Iraq before their deployment. But as he later recalled, "I did not have enough flying experience to fully learn advanced combat skills at that point. I was being exposed to them, but I was not really learning them. I was playing catch-up for a while."

In all, Bobby had two months' worth of flying with his troop before deploying to Iraq: up to fifteen flights each of up to two hours in length. By comparison, Tony Fuscellaro had the best part of a year to prepare with his troop and his squadron before deploying to Afghanistan. The problem was timing: Bobby was so keen to go to war that he had joined a squadron when it was already well into its preparations to deploy. The time before they departed for Iraq was very limited, but Bobby took advantage of every opportunity to learn, and by the time they arrived in Iraq he had just enough exposure to simulated combat to make him an effective member of the troop and the squadron.

In early June 2007 Bobby was on the squadron's advanced echelon flight to Kuwait, where he spent a month arranging hangar space and taking care of other preparatory tasks before the rest of the squadron arrived. He even got a bus driver's license to drive an old Iraqi bus. It was the beginning of two driving mishaps that earned him the good-natured nickname "Ricky Bobby," after the main character in *Talladega Nights*, a sports comedy about a NASCAR driver from West Virginia played by Will Ferrell.

Bobby's base was located in southeastern Mosul on the west bank of the Tigris River. Surrounded by Hesco barriers, the base was just under two miles long and half a mile wide. A U.S. infantry base twice the size of the air base was next door. The squadron flight line was about a quarter of a mile from the mess hall.

Pilot accommodations were Spartan but not bad by military standards in Iraq. Bobby and his fellow pilots lived in containerized housing units, large shipping containers insulated with wood panels. Each room had an air conditioner embedded in the window. Initially Bobby had to share the room, but when that pilot transferred to another base, he had the room to himself.

On the Fourth of July 2007, the harsh reality of war struck 4-6 Cav. The squadron was shocked to its core by the death of a beloved instructor pilot, CW2 Scott Oswell from Colorado Springs, Colorado, at the beginning of his second combat tour. On one of his first combat flights, he was killed when his helicopter hit overhead power lines near Mosul. His death struck Bobby hard. Chief Warrant Officer Oswell was an inspirational leader and teacher whom Bobby and the rest of the squadron loved for his personal warmth and selflessness. He and his wife had adopted several children and spent part of every weekend volunteering at a nearby homeless shelter. Bobby described Oswell as "an incredible human being, probably the best of us."

A day or so after Oswell's death, Bobby flew his first in-country patrol accompanying pilots from the 82nd Air Combat Brigade who the 4-6 Cav were replacing. During the patrol, he saw a horrifying incident on a street in Mosul. Insurgents captured a woman, bound her hands, and shot her in front of her children, leaving them sobbing over her dead body. It was an act of barbarous cruelty. They knew that the Kiowa team flying overhead would have to fly in low, get the position of the body, and then call it in so that the Iraqi Army could pick up the corpse. In other words, the insurgents used the murdered woman's body as bait so they could lure the Kiowa team into an ambush. They opened fire on the two Kiowas, but neither aircraft was hit. The Kiowas in turn fired a brace of rockets at the insurgents. There were no reports of any casualties on the ground. Observing the "incredible proficiency" of the pilots of the 82nd, Bobby thought to himself, "I have got a lot to learn."

Sitting in the trail helicopter, Bobby was also stunned. The entire incident had unfolded in less than thirty seconds. A young man with deep moral convictions, Bobby immediately saw that the insurgents were living in an alternate reality where these sorts of barbarous murders were acceptable. As he later recalled, "Their idea of moral behavior and mine were so far apart, I didn't see any way in which we could coexist." These were "bad people who needed not to be around anymore." Going forward, he would have no moral qualms whatever about killing insurgents and their Al-Qaeda allies.

Because Bobby had no prior combat experience, understandably, he lacked confidence. His second flight helped fix that and contributed to his development as a combat pilot. Once again, he was flying with a veteran of the 82nd now coming to the end of his tour. They were flying the south side of Mosul; their mission was to help a U.S. jet fighter overhead identify its target. The Kiowas were using smoke grenades, but the jet was flying too fast to see them. Bobby suggested to his copilot that they fire a rocket instead. After all, an exploding rocket would make a big smoke plume that the jet could see more easily. The way the copilot looked at Bobby indicated that he had not thought of this solution before. After a moment's thought the copilot agreed, as did the jet pilot. They fired a rocket and marked the target, and it was destroyed by a bomb from the jet. Bobby realized that he had a knack for improvising solutions.

CW3 Boise also helped improve Bobby's self-confidence. Bobby had bad experiences at West Point and at flight school when some professors expected him to know things he didn't because of the poor education he had received in West Virginia schools. Boise did not. When Bobby failed to answer one question on a gunnery skills course test that Boise had devised, Boise told Bobby that he could not be expected to know something if he had never been taught it. "Now you know it. You'll never forget it."

Bobby's flying schedule—like that of the rest of the squadron—was grueling: ten days on, one day off. And there was a Kiowa team on call twenty-four hours a day. They flew and fought around the clock. They had four shifts, each of about six hours. Pilots did not necessarily have to fly all six hours, but during that six-hour shift they were responsible for supporting the ground forces as needed.

The first shift began in the early morning just before sunrise and ended in midmorning. The second shift began at midmorning and ended in midafternoon. The third shift began in midafternoon and ended after sunset. The fourth shift was overnight.

The rhythm of each day was fairly consistent. The first shift was usually quiet, with little fighting going on. The Kiowas on duty might provide air cover for a ground patrol as they moved around Mosul. The second shift—known as the "jihadi shift"—was when all hell broke loose. That's when the ambushes and attacks on ground troops would begin and when IEDs would be set off. The third shift, right after sunset, is when the insurgents planted IEDs. At this time of day, the squadron's Kiowa patrols flew around looking for insurgents who were planting IEDs. If they found them, they killed them. The late-night shift was when U.S. troops conducted raids to capture insurgents and when those units needed close air support. If there were no raids, the Kiowas on duty flew peacefully over a sleeping city. Everyone in the squadron rotated through the four shifts, generally spending two to three weeks on each one. But Bobby preferred the most dangerous and "exciting" jihadi shift, and because he controlled his troop's flight schedule, he put himself on that shift more often than not. Captain Olson spotted him, however, and moved him to the deep night shift.

Throughout the squadron's deployment, it killed a remarkable number of insurgents as well as terrorists from Al-Qaeda in Iraq, the future ISIS. According to Iraq War scholar and author Michael Yon, who had access to Bobby's squadron intelligence officer, they killed 117 insurgents and terrorists between April 2007 and March 2008.[1]

This was attributable to not only the courage of the pilots but also their accuracy. One example Bobby recalls is of a fellow pilot shooting three rockets into a moving car. Lieutenant Colonel Jamison set the highest standards and expected everyone to achieve them. Indeed, he set out his expectations so clearly and effectively that the pilots in the squadron willingly took on extra practice sessions to improve their accuracy before every patrol. The desire to shoot well became a cultural expectation within the squadron rather than an order imposed from above by Jamison. The squadron had several practice ranges around Mosul and used them before beginning each patrol unless they were scrambling into the air to respond to an emergency call for help.

This commitment to greater accuracy was reinforced by the fact that they were operating in a densely populated urban environment where a lot of their rockets had to be fired "danger close." As Bobby recalled, "It was not uncommon for me to shoot rockets at a building on one side of the street with American soldiers on the other side." The fact that they were called upon to fire rockets daily in combat helped too.

The standard of accuracy that Bobby and his fellow pilots achieved was all the more impressive when you consider that their rockets were not designed as precision-strike weapons. Rather, they were area weapons designed to strike large targets such as supply depots from a mile or more away. In and around Mosul, Bobby and his fellow pilots often had to fire from such close range that they had to fly through the shrapnel of their own rocket hit.

Firing from close range was also necessary because the .50-caliber machine gun on Bobby's Kiowa did not work properly. Later he complained that "I never trusted it in a fight; the damn thing was jamming all the time."

Perhaps the main feature of the culture that Lieutenant Colonel Jamison built in the squadron was the need for "a continuous aviation presence," or "get out there and take the fight to the enemy." Jamison was not encouraging his pilots to fly for the heck of it. There was a lot of thought and experience behind this culture that dated back to Jamison's time as a young officer patrolling the Fulda Gap in the final years of the Cold War. At that time, he observed the deterrent value of a constant U.S. helicopter presence on the German-Czech border. This not only made East German and Czech border guards less likely to fire their weapons but also deterred intrusions into NATO airspace by East German and Czech aircraft.

Applied to the counterinsurgency war in Iraq in 2007–2008, Lieutenant Colonel Jamison's doctrine of continuous aviation presence was based on three pillars, all of them supported by hard evidence. The first was that intelligence reports proved that continuous aviation presence deterred insurgents. From his experience over the Fulda Gap, Jamison concluded that having a continuous aviation presence makes the enemy think that your capability is greater than it actually is. Classified intelligence gathered in Mosul told Jamison how much insurgent-planned activity did not occur because 4-6 Cav had aircraft up that day or night.

The second pillar of Lieutenant Colonel Jamison's doctrine was that a continuous aviation presence ensured a faster reaction time to insurgent activity and more rounds on target. As Jamison recalled, insurgents in Mosul and other Iraqi cities were very good at stopping a car at an intersection, getting a mortar tube out, setting it up, firing three rounds, putting the mortar tube back in the car, and driving off. They could plant IEDs in the ground almost as fast. So, by the time an unmanned aerial vehicle (UAV) picked up such activity and provided a video feed to the squadron tactical operations center (TOC) and a Kiowa team took off, it was too late. The incident was over, and the insurgents had disappeared.

The third pillar of the Jamison doctrine was that a continuous air presence emboldened U.S., coalition, and Iraqi ground forces in Mosul because it gave them more safe space to execute their mission.

Bobby was a strong advocate for Lieutenant Colonel Jamison's doc-trine, but there was resistance to it. One source of resistance was from warrant officers on their second or third tour who were tired of com-bat and frustrated by the tight rules of engagement. Another source of resistance came from brigade headquarters, where there was concern as to whether the squadron could maintain the aircraft when they were flying so many hours. Bobby, Captain Olson, SWO Boise, the senior instructor pilot, were strong advocates for the Jamison doctrine and helped their CO win over the rest of the pilots. Jamison himself con-vinced brigade headquarters that the squadron could maintain the air-craft even at this higher tempo of activity. And they did.

Lieutenant Colonel Jamison's strategy worked, but there was a price: an intense and exhausting number of flying hours per month. As Bobby recalled, "We were just flying our tails off." For him this meant that he always flew more than 100 hours a month. His average was 120 hours a month, and sometime he flew 130 hours per month. Army policy required a pilot flying this many hours to see a flight surgeon to check that he or she was not reaching the point of exhaustion. Jami-son put in place a system that required every pilot flying 100 a month and every 10 hours after that to be checked out by the flight surgeon. This was in response to brigade headquarters' concern about the strain Jamison's strategy was putting on his pilots. At the same time, brigade headquarters recognized that Jamison's approach was well thought out and driven by operational necessity.

The truth was that by the summer and fall of 2007, President George W. Bush's surge was fully under way, and the fighting in the city of Mosul was intense. Every day the Kiowas on the second and third shifts were in intense firefights. At this time, Bobby remembers the squadron operating as an airborne rapid-reaction force flying from one "troops in contact" engagement to another. And as terrorists planted IEDs every night, even the quieter fourth shift became more intense as the Kiowas did all they could to support the ground forces by killing the terrorists.

A typical day for Bobby during this period began with reconnaissance over a part of Mosul looking for insurgents or anything out of the ordinary. While he and a second Kiowa were doing that, they would get a radio call from a ground patrol that had just hit an IED and were taking small-arms fire. The Kiowas would respond immediately, providing whatever close air support was necessary. Normally when the Kiowas turned up, the shooting stopped. The Kiowas would provide security for the troops on the ground for ninety minutes before leaving to provide air cover for an army convoy. After a while of doing this, they would be called away to provide close air support for other troops in contact. Sometimes when the Kiowas turned up, the insurgents fired at them, and they had to fight back. Sometimes insurgents would fire on the Kiowas when they were flying a routine reconnaissance mission over the city. Within seconds they would be in a firefight. When Bobby was on the late afternoon or night shift, he flew around looking for insurgents on the street planting IEDs. On average, he encountered them twice a week and shot them with a missile.

By this point, Lieutenant Colonel Jamison had formed a clear impression of Bobby's capabilities years later. Jamison said, "I could tell he was intelligent, thoughtful, very deliberate, but not in a nerdy sort of way. He still had that aggressive spirit; he would take risks. And to have that combination of a person that's intelligent, thoughtful, and deliberate in the planning process but still able to go out there and push the edge, push the boundaries, that's a rare combination."

December 30, 2007, Distinguished Flying Cross

Bobby, his copilot Pete DeGorgio, and his trail Kiowa team Tom Boise and Susan Weathers were on the jihad shift, when the heaviest fighting normally took place. By Iraqi standards it was a cool day; the temperature hovered just below 60 degrees. Bobby was called to the TOC, where the staff were monitoring a UAV feed on a large television screen.

The two Kiowas on the early shift were tracking a van full of insurgents and a suspicious heavy weapon, but they could not fire because of the risk of killing or injuring innocent civilians.

The sight of a suspicious heavy weapon alarmed Lieutenant Colonel Jamison and his team in the squadron's TOC where the UAV feed was coming in. Jamison had received intelligence reports indicating that insurgents had acquired a ZPU 23-millimeter single-barreled heavy machine gun and were testing it for use against his squadron.

Based on an old Soviet-era design, the ZPU was primarily built to shoot down low-flying aircraft. Until now, the worst antiaircraft fire 4-6 Cav had to deal with were AK-47s and PKM belt-fed machine guns. A ZPU could fire up to 150 armor-piercing rounds per minute, which would enable the insurgents to threaten the safety of the squadron's aircraft and crew as never before. The ZPU had to be destroyed.

Bobby too was alarmed. He always tried to understand the bigger picture, so he studied the intelligence reports very carefully and had an unusually high level of battlefield awareness. As a result, he grasped how important it was to destroy the ZPU.

This was a tough mission. Once a Kiowa team was airborne, they had to track a target for hours through a large, densely populated city until it got to an area where there were no other buildings or civilians around and they could attack. The task was made more challenging because the sight was on top of the helicopter. The sight was the ball on top of the Kiowa that housed a thermal camera, a television camera, and the laser designator. It was put on top of the aircraft so it could hide behind a hill or clump of trees and spot tanks in conventional high-intensity warfare. But in densely populated cities such as Mosul, the location of the sight made it hard to see down into the streets. You had to look at an angle to see what you were looking at, and in a city such as Mosul, insurgents could easily drive a van or a car out of sight behind a building.

When Bobby, DeGorgio, Boise, and Weathers came back from their preflight check-in, the other Kiowa team was still tracking the van but had not yet had a chance to destroy it. They went back to the TOC, where they could see the van through the UAV feed. The team tracking the van radioed in that they were running low on fuel. Bobby was the mission commander and realized that he and his team had to take off immediately. If they didn't, by the time the other Kiowa team broke contact to fly back to refuel, it might be impossible to reacquire the van.

Bobby and his team took off and got the handover from the other Kiowas. As mission commander, Bobby was in the left seat, the primary tracker operating the sight. DeGorgio was in the right seat flying the helicopter. Boise and Weathers were in the trail helicopter. Bobby identified the van and took over the tracking.

After several hours of tracking, the insurgents drove their van into a junkyard on the western side of Mosul. The good news was that the insurgents and the van were in a place where there were no civilians and there was no risk of collateral damage. Under the rules of engagement, Bobby and his team were now free to destroy it. The bad news was that there were multiple big power lines in between the Kiowas and the van. It was a tricky shot, but Bobby wanted to take it. So, DeGorgio flew the Kiowa in between the power lines. Bobby lazed the target; Weathers fired a Hellfire missile, which Bobby guided toward the insurgents' van.

As the missile was inbound toward the van, a car entered the junkyard and pulled up alongside it. It was full of insurgents. As the missile got ever closer, the car moved a little away, and the Hellfire hit the van perfectly. Bobby had never felt such an adrenaline rush and such a sense of relief in any of his forty previous engagements. "All the pressure was on me to do this right," he recalled. "I thought 'if I mess this up, these people will get away and kill some Americans.'"

Bobby felt intense pressure to get this right. Once earlier in his deployment he had missed a target, and that failure weighed heavily on him. But this time, he was spot-on. As Bobby recalled, "I tracked the missile, and as soon as it hit, I felt a sense of euphoria. I did what I was supposed to do."

After the explosion, there was a big plume of smoke. Dust and debris flew everywhere, obscuring the Kiowa pilots' view of the target area. Both aircraft flew in a wide circle around the target area and switched roles.

Boise and Weathers were now in the lead, and Bobby and DeGorgio were in the trail aircraft.

As the smoke and debris began to clear, the UAV had spotted that although all the insurgents in the van had been killed, the insurgents in the car were now pulling any undamaged weapons from the back of the burnt-out van. They had to be stopped. The TOC cleared Bobby and DeGorgio to open fire with the .50-caliber machine gun they had on their Kiowa that day. They didn't want any more missiles or rockets fired into the target area because of the amount of smoke, dust, and debris they would put up in the air when they exploded. Bobby radioed Boise and Weathers to let them know that he and DeGorgio were attacking the insurgents with their .50-caliber machine gun. Bobby does not believe that Boise and Weathers heard that call because they continued to fly in circles around the target area.

Bobby and DeGorgio turned right, diving to attack the insurgents. They used a graveyard atop a rolling hill as cover against the insurgents and were only fifty feet above the gravestones. At that low altitude DeGorgio had to pull the nose up and then push it down, pointing it at the ground. Just as DeGorgio pulled the nose up over the crest of the hill, the squadron TOC radioed Bobby that insurgents were firing at him. "I can't hear it yet," he replied. "They're not getting close." But within seconds he heard a colossal volume of enemy gunfire going past the aircraft. It was louder than anything he had heard in his deployment to date.

DeGorgio climbed, turned the nose over, and fired the .50-caliber machine gun into the junkyard where the remaining insurgents still were. Bobby wasn't sure whether the insurgents stopped firing once DeGorgio opened up with the .50-caliber machine gun or whether the noise of the machine gun was so loud that it drowned out the noise of the gunfire below. But once DeGorgio stopped firing the .50-caliber machine gun, Bobby could hear the heavy gunfire again. Now they were getting close to the power lines, so DeGorgio made a hard left turn to avoid them. Just as he did so, Bobby heard the AK-47 and PKM machine-gun fire hitting the helicopter: "Tink, tink, tink, tonk."

Bobby wasn't worried about the "tink" sound. He had heard it many times before when his helicopter had been hit by small-arms fire. But the "tonk" sound was new. Later, Bobby remembered that "it felt like someone had hit the bottom of my helicopter with a sledge hammer. And I felt it reverberate through my body. Immediately, all the electronics in the helicopter shut off. The screens on the Kiowa's dashboard went blank. I couldn't hear anything because my radio went out. I looked over at Pete who was starting to autorotate the helicopter because he thought we had lost the engine. I told him to dump the collective quick."

Abruptly DeGorgio lowered the collective control, forcing the Kiowa into a fast descent and reducing the amount of power it needed from the engine. They both thought that the engine was failing and that they were going to crash. Right in front of them in the graveyard, Bobby could see a low two-foot wall around a family plot. He thought, "This is going to take my legs off because we are going to hit it, and it's going to go right through the bottom of this helicopter." Then the multifunction display, the screens on the Kiowa's instrument panel, began to reset itself, and the aircraft got half of its electronics back. So, when the redundant system reset itself, Bobby and DeGorgio could see that they still had engine power. With the communications system out, Bobby shouted to DeGorgio, "Hey man, we got power. Take off. Pull the pitch." DeGorgio was already doing this, and the Kiowa began lifting off.

As it did so, DeGorgio asked Bobby, "Should we land or go back to base?" As mission commander, Bobby looked around. He saw that they were far from friendly forces and close to the insurgents who were firing at their helicopter. So, Bobby said, "Let's go back to base." DeGorgio turned the helicopter around to return to base. As they flew, Bobby could smell a sickly sweet, smoking smell in the helicopter, suggesting a possible electrical fire. He tried calling Boise and Weathers, but they could not hear him because none of the four radios in the Kiowa could transmit. Bobby could, however, hear Boise and Weathers using the codeword "tumble," meaning "we've lost you."

As they flew back to base, Bobby was concerned that they were splitting the team, that Boise and Weathers were out there looking for them thinking they had crashed somewhere. To attract their attention, Bobby began to throw smoke grenades out of the helicopter. It worked. Weathers saw the smoke and recognized that they were in a rough line. Looking down the line, she saw Bobby and DeGorgio's Kiowa and concluded that they were heading back to base.

The route back to base took them over the "Tampa-Barracuda," one of the most dangerous parts of Mosul. Bobby began to experience a strange sensation: it felt like everything around him was moving slower than it seemed. He thought they were flying at forty knots. Later, he recalled that "I felt like we were just hanging there." Bobby shouted to DeGorgio "Faster, we've got to go faster. Man, I don't want to get shot down over 'Tampa-Barracuda.' We've got enough issues right now." DeGorgio just pointed to the air speed indicator, which showed one hundred knots.

They landed on the maintenance side of the Mosul airfield, standard practice when landing a shot-up helicopter, especially if you didn't know what was wrong with the aircraft. Because Bobby and DeGorgio were concerned that their Kiowa might catch fire, they parked it far from the flight line. Bobby was determined to complete the mission and told DeGorgio to shut down their damaged helicopter and that he would go and start up the squadron's spare Kiowa as quickly as possible.

Bobby grabbed his M4 rifle, his ammunition, and his water bottles and began running as fast as he could toward the spare helicopter, which sat about two hundred yards away. Once again, he experienced time distortion. He thought his legs were barely moving because they felt so heavy. He had spent so much energy in the engagement and there was so much adrenalin in his system. In reality, he was moving quickly. Corporal Crowley, one of the squadron's best mechanics, wanted to help. So, he took Bobby's rifle and ammunition and began running alongside him, cheering Bobby on. But despite this encouragement and his own determination, Bobby just could not get his legs to run any faster.

When Bobby reached the spare helicopter, he got in and started it. As its blades began to turn, the NCO from the unit that owned the UAV and who had been relaying information to him over the radio came down to the flight line to check that he and DeGorgio were okay. It was a tribute to the camaraderie that Lieutenant Colonel Jamison had created within the squadron.

As DeGorgio strapped in, Bobby saw Lieutenant Colonel Jamison coming across the flight line. Bobby's first thought was "I'm in trouble for getting the helicopter shot up. He won't allow me to take off." He had had helicopters shot up earlier. Jamison had admonished him about it. This time it was different. As he walked toward Bobby's Kiowa, Jamison had a huge grin on his face and gave him a big thumbs up. Later, Jamison recalled how proud of Bobby he was, but at the same time Bobby's action bothered him: "I felt a little bit of guilt because the month prior I had done the same thing. I thought to myself if something happens to that kid on that second flight that will be because he's only doing what his boss did. I probably set the wrong example for him."

Bobby and DeGorgio took off; the UAV had picked up the insurgents' car again. Guided by the UAV feed, Bobby and DeGorgio began to track it. The insurgents had been driving all over Mosul trying to lose the Kiowas. After an hour, they thought they had shaken off their American pursuers and made a run for it through the open desert. It was a fatal mistake. The spare Kiowa was equipped with high-explosive rockets, and Bobby was in the right seat ready to fire them. Boise and Weathers were in the trail helicopter equipped with rockets and one Hellfire missile. Weathers fired the Hellfire and hit the back side of the insurgents' car. The impact of the explosion caused the car to swerve and crash into a sand berm. The four surviving insurgents got out of the car and ran in three different directions through the open desert. The two Kiowas flew in a circle and engaged the insurgents with rockets before resuming their encircling pattern. Amid the plumes of smoke, dust, and flying debris, Bobby and DeGorgio were having difficulty keeping track of all of the insurgents. Bobby asked Boise over the radio whether he could see any of them. "No," Boise said, "but if you do, turn inside of me." Then in his peripheral vision, Bobby saw one of the insurgents running back toward Mosul. Bobby turned inside Boise and Weathers's helicopter and engaged the running insurgent with a rocket but fired too far in front of him and missed. Bobby readjusted his aim and prepared to fire again. This was a very difficult shot. Not only was the insurgent running, but the way the rocket exploded, it scattered its lethal shrapnel forward and to the right in a kind of butterfly pattern. So, with his last rocket, Bobby aimed low and to the left to ensure that the shrapnel spray would kill the insurgent.

Amazingly, Bobby's rocket hit the insurgent in the left thigh, killing him. It was a spectacular shot. Later with his usual modesty, Bobby attributed his success to luck rather than skill. As he flew a slow left turn, he asked DeGorgio to shoot the insurgent with his rifle. "Naw," DeGorgio replied. "You got him." Weathers killed another insurgent with her rifle. Firing at insurgents with rifles after rockets were fired was policy because the rockets normally didn't harm the enemy at all. It was therefore necessary to reengage with a rifle to ensure a successful result. The standard technique was to shoot the rockets first, then have the pilot in the left seat reengage if the insurgent was still combat-effective.

One insurgent got away, however, and hid in a crowd with other people. Under the laws of war and their rules of engagement, there was nothing Bobby and his team could do. And as a practical matter, they were out of ammunition anyway.

The Kiowas flew back to the Mosul airfield to rearm and then flew out again to see if this insurgent was now clearly visible and away from innocent people. Guided by the UAV feed, they found him again, but he was still hiding in a crowd of innocents and could not be attacked. It was now late in the day, and brigade headquarters decided to recall the two Kiowas. Ground forces would take over the monitoring. Later, U.S. soldiers captured the insurgent.

As they flew back to the Mosul airfield, Bobby was proud of what he had accomplished. But he was still worried that he might get in trouble with Lieutenant Colonel Jamison for getting his first helicopter so badly shot up and might even lose his pilot-in-command status. He needn't have worried.

Lieutenant Colonel Jamison and the rest of the squadron were so excited about what Bobby and his team had done and welcomed them home. After five hours of continuous engagement, Bobby and his team had eliminated a significant threat to the squadron and killed ten insurgents. This was a real tactical success in which as mission commander Bobby could take justified pride. Later, Jamison put Bobby's achievement in perspective. "It's one thing to do this in an Apache, but in a Kiowa with no armor, no sighting system for the rockets or the guns, just a grease pencil on the windscreen, you are back to World War II, to a P-51 Mustang trying to hit a moving train, getting down low, getting up close without the P-51's speed. You are really putting yourself at risk … [and] to successfully carry out this mission was extraordinary."

Lieutenant Colonel Jamison nominated Bobby and DeGorgio for the Distinguished Flying Cross and Boise and Weathers for the Air Medal for Valor. They were scheduled to receive them from Brig. Jessie Farrington, but Bobby was not there. He was lying in a hospital bed at the U.S. military medical center at Landstuhl, Germany, seriously wounded.

February 17, 2008: A Brush with Death

On February 17, 2008, Bobby was very seriously wounded. He was flying with Lee Russell in his Kiowa. Tony Galloni and Carlos Lopez were in the trail helicopter. Russell was a former special operations mechanic and was now a maintenance test pilot in his troop. The army was building a new combat outpost on the southwestern side of Mosul. Bobby and his team were providing air cover. An armored bulldozer excavating the site was under fire from a sniper. The driver was safe, protected by the armor, but the sniping had become a nuisance. Once Bobby and his team arrived overhead, the shooting stopped. The moment they flew away, the shooting resumed. So, they returned to cover the engineers. This happened three times. Finally, the engineers radioed Bobby with a solution to this cat-and-mouse game with the sniper: "Why don't you leave but don't go too far. The second he shoots, come back quick, and we'll use ourselves as bait." Bobby agreed.

At that moment, Bobby got a call over the radio from the squadron's TOC. They told him that a UAV had identified some insurgents on the north side of Mosul planting an IED near the junkyard where Bobby and his team had destroyed a van carrying a heavy weapon on December 30, 2007. Before leaving he checked that the engineers were okay, adding that he had to answer a priority call nearby. The engineers told him that the sniper fire had stopped and that they were now safe.

Bobby and his trailing Kiowa flew over to northern Mosul and began looking for the insurgents who had planted the IED. When they arrived, they could see four or five of them standing around next to a gas station. Thanks to the UAV feed, Bobby and his team had positive identification that these men were insurgents. The two Kiowas waited for them to leave the shelter of the gas station so they could kill them.

The insurgents must have seen the Kiowas, because all of a sudden all five took off running down the street. They were wearing dark blue tracksuits with a white stripe similar to those worn by Tony Soprano's enforcers in *The Sopranos*. The area was densely populated, so Bobby could not risk firing a rocket with so many innocent people around the insurgents. Then the five insurgents ran into a house. Bobby thought "Gotcha. We have you cornered." The two Kiowas began flying circles around the house so the insurgents could not escape down the street. Bobby also called U.S. soldiers so they could come and raid the house and capture the insurgents. Unfortunately, the ground patrol coming to capture the insurgents hit an IED.

Bobby and Russell had to remain on station because it was taking so long for another ground patrol to get to the house where the insurgents were holed up. Bobby and Russell were now running low on gas and needed to refuel. Luckily, Lieutenant Colonel Jamison and his trail helicopter were on patrol nearby, so Bobby briefed him and handed over the surveillance to him. As Bobby was away refueling, the second ground patrol reached the house and raided it. They only captured one insurgent. Either there was a tunnel underneath the house through which the other insurgents escaped or they had slipped away down the street.

When Bobby and Russell came back on station, the ground troops were down at the gas station trying to gather more information about the missing insurgents. The troops were holding a suspect dressed in a dark blue tracksuit with a white stripe. Tracksuits of various colors had become a kind of uniform for Iraqi insurgents. But the blue one with a stripe was unique. The ground troops wanted Bobby to confirm that the man was one of the five insurgents he had earlier chased down the street.

Bobby told the ground forces that he could not see the suspect's face from the air and that it would be hard to confirm his identity. Reluctantly, Bobby agreed to take a look. So, he and Russell took their Kiowa down low, flying lower and slower than they normally would so Bobby could get a good look at the suspect. When the soldiers pulled him out of the back of a Bradley fighting vehicle, Bobby could see that it was the same man wearing a dark blue tracksuit with a white stripe like one of the people he had been chasing earlier. He told the ground forces so they could check him for explosive residue. As a result of that test, they were sure he was the insurgent who had planted the IED.

At this point, Bobby and Russell were only twenty feet off the ground flying at between thirty and forty knots. They were now a much easier target for insurgents on the ground. As a result, once Bobby had completed the identification, they began to speed up to gather more air speed and to minimize the risk of being shot. Just as they began to increase speed and gain altitude, Bobby looked down and saw two Iraqi buses parked face to face down the street from the gas station. He saw Iraqis running in all different directions beneath his Kiowa. This was normally a sign that there was gunfire on the street that Bobby and Russell could not pinpoint. The sound of the gunfire was reverberating off the walls of the buildings, so no one knew where it was coming from. Sitting in the left seat of his helicopter, Bobby reached up for his rifle and stepped on the mike switch to let the trail aircraft know there was gunfire in the street below.

As Bobby did so, two enemy machine guns on opposite sides of the street fired into Bobby and Russell's helicopter, hitting both sides of the aircraft. One of the rounds came up underneath Bobby's seat through the heater duct into the back of his left knee and out the front and then lodged in the instrument panel in front of him. Bobby grunted in pain. The impact of the machine-gun bullet hitting his knee felt like it had been hit by a baseball bat. The next sensation was a numbing feeling.

Bobby knew he had been shot. He put his rifle back on the Kiowa's instrument panel and stuck his hand down his left leg to feel for the wound. When he looked at his hand, it appeared to be covered in red paint. In combat, there are three places you don't want to get shot: a bone, the nerves, or an artery. Bobby was hit in two of the three: a nerve and an artery. The large amount of red blood he was seeing was the result of his artery being nicked.

At that moment Russell, his copilot, had pulled all the power the helicopter had. The helicopter's warning system told them they were pulling too much power. Bobby thought, "Oh man, this is going from bad to worse. I've been shot and we're about to crash." Russell got the helicopter under control, however, and turned it toward the U.S. military hospital in Mosul. Luckily, it was less than four minutes' flying time away.

As the Kiowa flew to the hospital Bobby tried to get his tourniquet off his belt, but it got caught on the pouch's button. The moment he let go of his left leg to pull the tourniquet out, blood began squirting out of his leg from the severed artery. Bobby put his left hand back on his wound to stop the bleeding. He told Russell he could not get the tourniquet out. Russell said, "Don't worry. We're almost there." Bobby held his left leg with both hands as tightly as he could to stem the blood flow. As he did so, he began to feel faint from the blood loss he had already sustained and thought he was going to pass out.

Russell patted Bobby's right thigh and said, "Stay with me, man; stay with me, man. We're almost there." Bobby, a little irritated by Russell's well-intended words, replied, "I'm fine. Just fly the helicopter." But what Russell had said gave Bobby a boost of adrenalin. He remembers thinking, "I'm going to live. If I can avoid bleeding out before, I'll be fine when I get to the hospital because they will give me a blood transfusion."

Russell shut down the helicopter when it landed on the hospital helipad and immediately applied a tourniquet just above Bobby's wound. Bobby took off his flight helmet. They were both puzzled that no medics came running out to meet their Kiowa as they normally would whenever a medevac helicopter landed. After a few minutes, an Army doctor put his head around the helipad barrier. Immediately, Russell started shouting at the doctor demanding to know why no medics were coming to the helicopter. Using the most colorful language he could think of—and he had quite a vocabulary—Russell demanded that the doctor come to their Kiowa immediately.

It was all a misunderstanding. The hospital had been told to expect a medevac helicopter with a wounded pilot. Usually this was a Black Hawk with a red cross insignia on the side. So, when a Kiowa landed unexpectedly, at first they did not know what to make of it. Once they understood that the wounded pilot they had been expecting was in the Kiowa, the doctor and his medical team brought Bobby inside.

Immediately, they applied a tourniquet, gave Bobby a blood transfusion, and took an X-ray of his left leg. Once the doctor read the X-ray, he told Bobby "Great news. The bullet nicked an artery, but it looks like it's closed up. The bullet hit a nerve bundle, but that will heal up over time. We're not going to send you home. Rest around the base, and you'll be flying again in a month or two."

The surgeons had to remove some shrapnel from the helicopter that the bullet dragged into Bobby's leg. The bullet itself had fragmented inside the leg, so the surgeons had to remove this shrapnel too. Once Bobby awoke from surgery, he had an O_2 device on his finger measuring his oxygen level.

Suddenly, the machine to which the O_2 device was attached started pinging loudly. A nurse came over, adjusted the O_2 device, and checked the machine, which was now pinging even more loudly. To Bobby, the nurse looked perplexed. She brought over a doctor to determine what was wrong and what to do.

After two chest X-rays, the medical team determined that a small piece of shrapnel had gotten into a vein in Bobby's left leg and had traveled up through his heart and into his lungs causing a saddle embolism, which blocks the blood flow to both lungs. This was potentially fatal because it cut off the blood flow to his lungs, so his oxygen level was crashing. His medical team pumped Bobby full of blood thinners to treat the clot in his lung. The problem was, that shrapnel was sharp and was still moving around in the artery in his lung.

The medical team now faced an urgent question: Should they try to remove it surgically or just leave it, hoping it would embed itself in the wall of the artery? Their difficulty was compounded by the fact that they lacked the specialized surgical equipment to remove the shrapnel and were concerned about the dangers in trying to move Bobby out of Iraq.

The medical team consulted a leading specialist at Walter Reed Army Medical Center in Washington, D.C., one of the top experts in his field in the world. His advice was to stabilize Bobby and move him to Walter Reed as soon as possible. Bobby's doctors had planned to send Bobby out that night, but for some reason the aircraft did not turn up. So, for that night Bobby was carefully secured to prevent him from moving lest the sharp piece of shrapnel in the artery in his lung move and cause even more serious damage.

The next morning, the Mosul medical team stabilized Bobby and carefully loaded him aboard an Air Force C-130 bound for the U.S. military medical center in Landstuhl, Germany. Not long after Bobby landed there, Lt. Gen. David McKiernan, commander of U.S. Forces Europe, invested a heavily medicated Bobby with his Distinguished Flying Cross. Bobby was so heavily drugged by the doctors that he barely remembers Lieutenant General McKiernan giving him his medal.

Medically, Bobby's problems grew more complicated. The blood thinners given in Mosul and Landstuhl eliminated the blood clot in his lung as well as the little clots in his left leg. But now he had compartment syndrome in the left leg: there was too much blood leaking into the muscle compartments, and it cut off circulation to his left foot. To treat it, the surgeons at Landstuhl had to give Bobby a fasciotomy, a cut from his left knee to his left ankle through the muscle compartments to relieve the pressure.

Bobby woke up from surgery with an alarmed nurse slapping him on the face saying, "Stay with me, stay with me." After surgery, Bobby had bled much more than they had expected, and his blood pressure was crashing. The Landstuhl team stopped the bleeding and gave him a transfusion. Now that he was stabilized, they prepared to fly him to Walter Reed in Washington, D.C.

Heavily medicated, Bobby was in intensive care at Walter Reed for several weeks. His sister met the evacuation flight from Germany. His mother and father drove down from West Virginia immediately afterward. They stayed in Washington, D.C., for several weeks, visiting Bobby daily, giving him all the loving support they could.

At first, the specialists at Walter Reed were unsure whether or not to operate to remove the piece of shrapnel from the artery in his lung. They moved Bobby to the pulmonary department of the hospital where, after careful analysis and deliberation, they decided not to operate. Their reasoning was that Bobby had already experienced so much trauma in his leg that more trauma in his chest would only make matters worse. Moreover, to operate to remove the shrapnel would have meant cutting away two of the five lobes of his lung just to get to the shrapnel in the artery. This was all too risky especially when there was a good chance that the piece of shrapnel would calcify and become part of the wall of the artery. It did. Today, Bobby carries that piece of shrapnel in his lungs. It doesn't bother him unless he gets a chest cold or laughs too heartily.

Bobby's convalescence in Walter Reed was slow and psychologically difficult. He had been ripped out of combat: one minute he was engaged with the enemy, and five minutes later he was in the hospital seriously wounded. Within days he was in Walter Reed. He experienced none of the normal transition from combat zone to home: no winddown period, no time in a staging area on the way home, no classes to facilitate homecoming. It was wrenching.

For Bobby, the difficulties were compounded by the painkillers and blood thinners he was on. "My experience is that I felt crushing guilt that I had left my troop," he recalled. "I remember at the time thinking that it was illogical, it was pure emotion, but then I couldn't help it. I couldn't get over it. I remember lying in the hospital bed at Walter Reed just emotionally drained from the guilt." Not being allowed to go back to Iraq because of the medication he was on really bothered him, and he had difficulties dealing with that.

Bobby left Walter Reed as soon as he possibly could. The doctors had told him that he was going to be there for another six months, so Bobby began a campaign to convince the doctors that he could complete his convalescence back at his unit's home base at Fort Lewis, Washington. Despite the excellent medical care, he found Walter Reed so depressing and was determined to leave.

The problem was that Bobby was in Walter Reed at the height of President Bush's surge, and the hospital was full of seriously wounded soldiers. Bobby was on a floor with soldiers who had lost one or both of their legs. Some were in good spirits and showed remarkable resilience. Bobby found them encouraging and uplifting. But understandably, there were many others dealing with the emotion and trauma of their injuries. Bobby would see them arguing with their wives in the hospital corridor. The traumatic brain injury unit was nearby, and Bobby found it "really crushing" to meet senior NCOs and officers who had been so badly injured by explosions that they could not carry on a conversation.

The Wounded Warrior Project helped Bobby in ways that he didn't even know he needed. When he was an inpatient at Walter Reed, every Thursday evening a Wounded Warrior Project volunteer would bring him a milkshake. When he was an outpatient, the same volunteer would take him out for a steak dinner once a week. Because Bobby spent so much time alone, just being able to be out and around other people was so positive and so important.

During this difficult time, what also really helped Bobby were visits from his brigade commander, Col. Jessie Farrington, and squadron commander, Lieutenant Colonel Jamison, when they were home on leave. Both senior officers improved Bobby's morale by helping him focus on how to prepare for the next stage of his army aviation career: a transition to Apaches. Both of them said they would help him make that transition, explaining that the army was going to phase out the Kiowa and that he should switch to Apaches now. Bobby loved the Kiowa and its scout mission, so initially he resisted. Jamison told him frankly, "This is an opportunity you cannot pass on." Bobby didn't.

During this difficult time, Bobby did not experience PTSD, anxiety attacks, or nightmares, but as he recalled, "I felt guilty and kind of useless; I was just hanging out trying to get better while other people were off doing important things. That was my challenge. I was singularly focused on getting back in combat. Emotionally, that is what drove me. I needed to get back in the fight." He only felt himself again when, in 2009, he redeployed into combat in Afghanistan. Being back in a combat environment was genuinely therapeutic.

While Bobby was at the Walter Reed Army Medical Center, the only thing he wanted—and wanted badly—was to return to combat with his squadron in Mosul. On one occasion a major general visited him and asked, "Is there anything I can do for you, son?" Bobby replied that the only thing he wanted was to get back to his unit in Iraq. "I want to be back in Iraq with my guys." The major general looked at him for a moment and said, "We'll see what we can do." About fifteen minutes later, the surgeon on duty came in and gave Bobby a tough talking-to. "Listen here. Nobody is going to let you go back to combat. You're on blood thinners. The only person allowed to let you go back to combat is me, and I refuse to do that because it would endanger your life. Asking to go back to combat is just making things worse for everybody." Bobby apologized.

In the early summer of 2008, Bobby returned to Fort Lewis, Washington, to complete his recovery and rejoin his squadron when they came back from Iraq.

23

STEPHEN TANGEN

Afghanistan, 2010

Joining the Storied 101st Airborne Division

After graduating from Ranger School in March 2009, Stephen Tangen went home for leave before joining the storied 101st Airborne Division at Fort Campbell, Kentucky. Because of his high class rank at West Point, Stephen had a number of assignment choices. But he chose the 101st Airborne for three reasons. The first was personal. Fort Campbell was closest to home in Naperville, Illinois. He wanted to remain connected to his beloved parents, grandmother, and brothers and to his girlfriend, Melissa, whom he would later marry. The second reason was his deep admiration and respect for the history of the 101st. This began when Stephen had watched the Tom Hanks/Stephen Spielberg miniseries *Band of Brothers*, which tells the story of Maj. Dick Winters and the men of Easy Company from the formation of the 101st in 1942 through D-Day, the Battle of the Bulge, and their capture of Adolf Hitler's mountain retreat in the Bavarian Alps. Stephen was inspired by

Dick Winters's selfless courage, inspirational leadership, and devotion to his soldiers. The third reason Stephen chose the 101st was its unit pride. As he later recalled, "Wherever you go in the Army, if someone has been in the 101st, more likely than not they are wearing the 101st sleeve patch. There is a lot of pride in that patch. That's what pulled me towards the 101st."

On April 20, 2009, Stephen reported to Fort Campbell along with his West Point classmate and roommate from Fort Benning, Dan Konopa. They joined "No Slack," the 2nd Battalion, 327th Infantry Regiment, 1st Brigade Combat Team, 101st Airborne. The battalion commander, Lt. Col. Joel "JB" Vowell, welcomed them. Tall and thin with striking blue eyes, brown hair, a wide, warm, welcoming smile, and a brilliant mind, Lieutenant Colonel Vowell was a decorated combat veteran. Commissioned from the University of Alabama through the Reserve Officers' Training Corps in 1991, he had earned a bachelor's degree and two master's degrees. He had also earned a reputation as a courageous officer, a serious defense intellectual, and an inspirational leader in the mold of Gen. David Petraeus. Long before 9/11, Lieutenant Colonel Vowell had presciently predicted that the next war the U.S. Army would have to fight would be a counterinsurgency conflict against a dispersed enemy embedded among its own people, not a conventional war against a state adversary such as the Soviet Union.

With his typical flair, Lieutenant Colonel Vowell welcomed Stephen and Konopa to No Slack, adding "this is the best damn battalion in the Army. Why?" Puzzled, Stephen and Konopa looked at each other. "Because you're here," Vowell responded. The very next morning, Vowell took Stephen and Konopa for a run. Stephen was still experiencing some joint pain from Ranger School, where he had pushed himself to the outer limits of his physical endurance. Vowell was superfit: a remarkable runner and cyclist. As they ran at an ever-increasing pace, he asked the two young West Pointers whether they had read Anton Myrer's classic 1,300-page military novel *Once an Eagle*, described by former chair of the Joint Chiefs of Staff Gen. Martin Dempsey as "simply the best work of fiction on leadership in print." Konopa replied that he had read it as a cadet. Stephen said "no." Vowell told Konopa he had two days to write a book report on Myrer's book. He gave Stephen two weeks to read the novel and turn in his book report. Vowell gradually increased the tempo, eventually leaving both young lieutenants gasping in his wake.

Stephen was assigned to the 1st Platoon, Charlie Company, and Konopa was assigned to Delta Company. Later, Stephen walked into his platoon office and found a handwritten note from Nick Eslinger, the departing platoon leader. Nick had just returned with the platoon from Iraq, where, as we have seen, he had dived on a grenade to save his men. In his note, Nick wrote that being a rifle platoon leader was an awesome opportunity, that the platoon had fought well in Iraq, and that they were the best platoon in the battalion. "You are very fortunate, you are going to do great, and I'll see you when I get back in a few weeks." When Nick got back from his ranger assessment he was busy packing up, but there was time for the two to catch up and for Nick to brief Stephen on the platoon.

Assuming Responsibility

Now it was Stephen's turn to fill Nick's shoes: take over the 1st Platoon and build it into an efficient, motivated fighting force. To that end, Stephen had to build that mutual devotion and trust so vital in a rifle platoon at war. In his Vietnam War classic *A Rumor of War*, Philip Caputo describes this best as "the intimacy of life in infantry battalions, where the communion between men is as profound as any between lovers. Actually, it is more so. It does not demand for its sustenance the reciprocity, the pledges of affection, the endless reassurances required by the love of men and women. It is, unlike marriage, a bond that cannot be broken by a word, by boredom or divorce, or by anything other than death. . . . Such devotion, simple and selfless, the sentiment of belonging to each other was the one decent thing we found in a conflict."[1]

That is the intimate bond that binds an infantry battalion. Stephen had a year to build it with the men of the 1st platoon, Charlie Company, before their next deployment.

Building that bond was difficult: the 1st Platoon was in transition. When Stephen assumed responsibility, some members of the platoon were processing out of the unit or the Army. Others were combat veterans skeptical of a twenty-three-year-old from West Point who had never seen combat. Others still were brand-new recruits. Above all, there was the legend of Nick Eslinger. At first, Stephen knew nothing of this extraordinary story. The first he learned of it was from the platoon NCOs. "You've got big shoes to fill, Lieutenant," they said. "What are you guys talking about," Stephen replied? It was only then that he learned about Nick's act of selfless courage and understood the true scale of the challenge he himself now faced.

But this young West Pointer had real assets. First, Stephen was an outstanding young leader who had received the best education and training possible. He had thought long and hard about what it takes to lead and motivate soldiers. Second, he loved soldiers and NCOs, and it showed. His soldiers could see before them a young officer deeply committed to their well-being, an officer who would always put their needs ahead of his own. Third, he was mentally and physically tough: a recent graduate of Ranger School with the Ranger Tab on his sleeve for all to see. It was the first thing everyone from Lieutenant Colonel Vowell to his platoon NCOs and soldiers looked for. It gave them confidence that Stephen had the mental and physical toughness necessary to cope with the unique pressures of fighting a war in Afghanistan, where the enemy was everywhere and it was often impossible to distinguish innocent civilians from combatants.

Stephen's approach to his soldiers was powerfully shaped by his West Point experience. At the USMA, Stephen had to know the underclassmen he was responsible for. He had to learn how to influence, motivate, and improve their performance. Now at Fort Campbell, he had to apply what he had learned at West Point and Ranger School. "You have to get to know them," Stephen said. "You have to know where they're from, know about their families, and know what they want to accomplish. Another thing that really connects with them and helps them build trust in you is when you surprise them and call them by their first name in private. You find an opportunity to just walk up to somebody and say 'Hey Eric, how are you doing?' The lights in his platoon office were on before dawn and long after dusk; his door was always open."

Stephen also set high standards of physical fitness. He knew that he had to be the best runner, the "best push-upper." Among his soldiers, he was known as "the guy who ran them to death." They tried to beat him but never could. Stephen always handled this with a light touch. At the end of a run he often told soldiers who could not keep up with them, "It's okay, man. Maybe next time."

Building trust also resulted from sharing his soldiers' hardships, eating what they ate, and sleeping on the ground during exercises just as they did. However close the bond Stephen established with his platoon before deploying, he was emphatic about the need for clear moral and ethical standards. As he later recalled, "If your soldiers know there's even a fraction of a percent's worth of wiggle room on the moral and ethical scale, the bad ones will take advantage of it. You must never tolerate unethical behavior in your soldiers. It is the right thing to do, but it is not the easy thing to do." In that year, Stephen built the foundation for a highly effective close-knit platoon. His soldiers were focused and highly motivated. Everyone knew what the platoon was about and what they needed to do, and they were passionate about it.

Stephen could not have accomplished this without the friendship and mentorship of SSgt. Eric Shaw, who later was killed in action by the Taliban during Operation Strong Eagle I. A native of Maine, a graduate of the University of Southern Maine in Gorham, Staff Sergeant Shaw was a big man with thinning brown hair, a broad face, a warm playful smile, and a great sense of humor. He was a kind, selfless man who always put others first, a perfect complement to Stephen. Above all, Shaw was a combat veteran of the 101st who had served two tours in Iraq. Every morning, he was in the platoon office thirty seconds ahead of Stephen just to tease him. Together with Cpt. Juan Garcia and 1st Sgt. Timothy Malmin, they shaped the platoon into an effective fighting force.

There were, of course, tensions and difficulties along the way. Some of Stephen's soldiers and even Stephen himself had to learn some hard lessons. For him it was humility. He had motivated and trained his platoon to be the best in Charlie Company. He had also instilled pride in his soldiers. But he was a little too outspoken about it outside his platoon. A good NCO in one of the other platoons told him to stop. After that, Stephen did stop.

The proof of the effectiveness of Stephen's leadership came at the JRTC at Fort Polk, Louisiana, in January 2010. There, units rotate through for a month prior to deployment. The JRTC is the U.S. Army's leading combat training center, a rigorous testing ground for any unit before going into harm's way. The JRTC's mission is to improve each unit's effectiveness as a fighting force. To that end, its exercises simulate real warfighting as realistically as possible. The exercises are not tightly scripted scenarios but instead are fluid, fast-moving battles in which the opposing force plays a smart enemy who anticipates your unit's moves and countermoves. Officers, NCOs, and soldiers have to adapt quickly and think fast. No mistakes go unnoticed. At the JRTC, the performance of Stephen's platoon proved decisive for their company and battalion in winning their exercise.

Kunar Valley, Afghanistan: Operation Strong Eagle I

No Slack arrived in Afghanistan in late April 2010 and began to take over responsibility for security operations in the Kunar Valley from the 2nd Battalion, 503rd Infantry Regiment, 173rd Airborne Brigade Combat Team.

Kunar Province is in northeastern Afghanistan. To the east is Pakistan, to the west is Laghman Province, to the south is Nangarhar Province, and to the north is Nuristan Province. Kunar's physical beauty is stunning. Its landscape can remind you of parts of Colorado, with rocky mountains over a mile high, deep narrow valleys, and sweet-smelling Alpine forests. In fact, nearly 90 percent of Kunar Province is mountainous.

The Kunar River valley dominates the province and bisects the range of mountains known as the Hindu Raj, the southern portion of the Hindu Kush mountains. The Kunar River cuts a narrow sixty-mile-long swath through Nangarhar and Kunar Provinces before flowing over the border to Pakistan. The Ghaki Valley links the Kunar River valley to Pakistan through the Ghaki Pass. On the floor of the Ghaki Valley was a stony one lane dirt road bordered on the north and south by steep ridgelines nearly a mile high. This was a very difficult land-scape for Stephen and his fellow soldiers to operate in.[2]

The population of the Ghaki Valley was insular, deeply conserva-tive, and largely illiterate. They were poor Mashud Pashtun farmers who raised crops (including poppies) and were as suspicious of Afghan officials from Kabul as they were of American and British aid workers, soldiers, and diplomats.

The Ghaki Valley was strategically important to the Afghan national government and to the U.S. and coalition forces. For the Afghan govern-ment and the Kunar provincial government, control of the Ghaki Valley was essential to enable it to connect with the people there and estab-lish political legitimacy with them. In security terms, Kunar Province had long been a safe haven for not only the Taliban but also terrorist groups including Al-Qaeda and Lashkar-e-Taiba. Controlling the Ghaki Valley would not only give the Afghan National Army and U.S. forces freedom of movement up to the Pakistani border but would also close off an important infiltration route for Taliban fighters coming in from Pakistan as well as shut down their escape route at the end of the annual fighting season just before winter.[3]

Operation Strong Eagle I was the latest and largest U.S. military operation in eastern Kunar Province since the Anglo-American inva-sion of Afghanistan in 2001. For the first two years afterward, the U.S. military presence was minimal, principally special forces conducting medical and civic assistance programs and counterterrorism operations. From November 2003 to June 2006, the U.S. Marine Corps stationed one battalion in Kunar Province on a seven-month rotation. It was only in April 2006 that the U.S. Army introduced larger numbers of conventional forces in permanent bases in Kunar Province. Two battalions from the 3rd Brigade Combat Team of the 10th Mountain Division accompanied by most of a battalion of U.S. Marines from the 1st Battalion, 3rd Marines, launched counterinsurgency operations in Korengal Valley in Operation Mountain Lion. Afterward the Marines returned to the United States, and the two Army battalions remained.[4] As John McGrath of the U.S. Army Combat Studies Institute explains, "From June 2006 to January 2009, this two-battalion force . . . conducted counterinsurgency operations throughout Kunar Province. . . . In order to increase the coalition presence along the Pakistani border area, commanders in the Coalition added another maneuver battalion to the forces in northeastern Afghanistan . . . in January 2009."[5]

Later that year, the Army presence increased to a full brigade (the 4th Brigade Combat Team, 4th Infantry Division), but as the Taliban insurgency increased, the four maneuver battalions were stretched thin. When Stephen's parent unit, the 1st Brigade Combat Team, took over responsibility in early 2010, the Taliban controlled the Kunar River valley.[6]

When No Slack arrived in Afghanistan in late April 2010, the battalion took over responsibility for security operations in the Kunar Valley from the 2nd Battalion, 503rd Infantry regiment, 173rd Airborne Brigade Combat Team. The No Slack troops were based at FOB Joyce on the east side of the Kunar River. Nestled at the foot of the Hindu Kush mountains, FOB Joyce was a hundred-acre fortified compound, named in honor of Marine LCpl. Steven Joyce, killed there on June 25, 2005. Approximately seven miles to the east was the border with Pakistan, and eight miles to the west was the Korengal Valley, just abandoned that month after fierce fighting and heavy American casualties. A few miles to the south was the Tora Bora cave complex where Osama bin Laden had escaped U.S. special forces in 2001.

The Taliban attacked FOB Joyce daily, but it was well protected, with hundreds of Hesco barriers, zinc-aluminum–coated steel mesh containers filled with soil, sand, gravel, or pieces of rock forming an effective outer perimeter. There were also plenty of armored watch towers to direct counterfire. This offered good protection against RPGs, recoilless rifle barrages, AK-47 fire, and PKM machine-gun fire. Buildings within the base were well protected by reinforced concrete roofs topped by sandbags, a big improvement from the canvas tents that No Slack occupied when it arrived. The base had a helicopter landing pad that could handle Kiowa Warrior, Apache Attack, Black Hawk, and Chinook troop-carrying helicopters and also provided a forward arming and refueling point.

On arrival in Afghanistan, Stephen's platoon (along with platoons from A and B Companies) were detached from Charlie Company to the battalion's headquarters and headquarters company (HHC) to provide the battalion with an additional maneuver company, that it would not normally have. The HHC needed Stephen's platoon to take responsibility for the Marawara District, the most populated town in the Ghaki Valley.[7] The village of Marawara, the district center, was a relatively secure place near the banks of the Kunar River. As the planning process for Operation Strong Eagle I was going forward at the brigade and division levels, Stephen and his platoon focused on the political outreach with the district governor and the local elders. For Stephen, it was interesting and rewarding. For his soldiers, it was monotonous. Pfc. Eric Soriano, a small lad from California, the gunner in Stephen's MRAP armored personnel carrier, sometimes questioned why the battalion was there. Private Soriano sat outside the buildings where Stephen's meetings took place in his turret with his .50-caliber machine guns scanning the high ridgelines for possible Taliban gunmen who might interrupt the meetings with hostile fire. Stephen understood Soriano's frustration. Other platoons had engaged the enemy, but Soriano's had not, and the question of what did we train for frequently arose.

As the weather warmed and the snow melted in the nearby mountain passes between Afghanistan and Pakistan, heavily armed Taliban fighters began to cross the border. The number of attacks on U.S. forces in Kunar Province increased dramatically, far above the level of 2008 and 2009. Within days of arrival, the battalion had begun taking casualties.

The worst incident occurred just four weeks after they arrived in Afghanistan. A massive IED exploded under a Humvee, killing five soldiers. It blew them to pieces, leaving body parts as well as parts of the truck scattered in every direction. Stephen was crushed, shocked by the horrific violence involved. He knew one of the victims personally: A Company's first sergeant, Robert Barton, a charismatic, driven leader. The five deaths also shocked the men of Stephen's platoon who, unlike their fellow soldiers in other platoons and other companies, had experienced no contact with the enemy to this point. Like a gut punch, the harsh reality of war in the Ghaki Valley had begun to hit home.

Stephen and his platoon were understandably anxious. Their anxiety was heightened because they were aware of the grim fate of the 334th Soviet Spetnaz battalion ambushed in the Ghaki Valley in April 1985, which had suffered 90 percent casualties. Anxious and restless though Stephen and his platoon were, they were also ready and eager to fight the Taliban.

To Lieutenant Colonel Vowell, it was clear that the threat emanated from the towns within the Ghaki Valley: Sangam and especially Daridam were Taliban strongholds. To deal with that threat meant fighting one of the Taliban's most ruthless and effective regional commanders, Qari Zia Rahman, who led Taliban forces in the Ghaki Valley. He was also a brigade commander in Al-Qaeda's elite "Shadow Army."[8]

Born in the late 1980s and raised in Barawolo Kalay, a village in Marawara District, Ghaki Valley, Rahman was a Mohmand Pashtun like the rest of the population. With such strong local roots, Rahman was popular in the valley and played a key role in helping the Taliban establish their legitimacy there. Rahman also doubled as an Al-Qaeda leader in northeastern Afghanistan and across the border in the Pakistan tribal areas of Bajaur and Mohmand. In the spring of 2010, Rahman commanded about three hundred insurgents, mostly Mohmand

Pashtuns from the valley but also from other parts of Afghanistan, Pakistan, Chechnya, Uzbekistan, Turkmenistan, and some Arab countries.[9]

Rahman was also a master recruiter and teacher of insurgents. He was "responsible for establishing training camps ... used to indoctrinate and train females, including children, to carry out suicide attacks on both sides of the Afghan-Pakistan border" and "for a female suicide attack in Kunar province on June 21, 2010 that killed two US soldiers."[10] In an interview for the documentary *No Greater Love*, Lieutenant Colonel Vowell summed Rahman up as "one of those who had helped train Bin Laden, a total sociopath, Sopranesque, truly a dangerous man."[11] Vowell saw Rahman's stronghold as a threat that had to be eliminated. Just days after he arrived at FOB Joyce, Vowell asked for, and received permission from brigade commander Col. Andrew Poppas to begin planning a battalion-level operation to break Rahman's hold on the Ghaki Valley.[12]

Strong Eagle I: The Plan

Lieutenant Colonel Vowell's plan was thorough and carefully choreographed. His overarching objective was to surprise Rahman and force him and his Taliban troops to stand and fight. The initial goal was to capture the village of Sangam, which lay between the U.S. and Taliban-held parts of the Ghaki Valley. The next goal was to move farther east and seize the Taliban stronghold of Daridam.[13] To this end, Vowell's military choreography was impeccable. As John McGrath has written,

> A platoon each from A and B Companies would be flown by helicopter under the cover of darkness to occupy dominating positions to the north and south of the valley. The A Company force, codenamed Team Gator and coming from FOB Monti located near Asmar about 18 miles north of Asadabad on the Kunar River, would occupy LZ Owl (later OP Shaw). The B Company platoon, designated Team Bayonet and reinforced with

the company tactical command post, was to come in from the southern Kunar Valley, and occupy LZ Hen (later OP Thomas). These positions were between 1000 and 1500 meters from the valley floor. An important part of the plan was that these forces had sufficient firepower with them. The platoons, accompanied by ANA troops, would carry heavy weapons with them— M2 .50-caliber machine guns and MK19 grenade launchers. Vowell hoped to use the two OPs to deny the enemy the ability to emplace heavy machine guns to fire down onto the troops advancing in the valley.[14]

The central part of Lieutenant Colonel Vowell' plan was a ground attack up the Ghaki Road commanded by Capt. Steven Weber, with Stephen's platoon in the lead. Its goal was to capture and clear the Taliban stronghold of Daridam. Afghan forces would lead. Protecting Stephen's left flank was a platoon led by Lt. Douglas Jones. His mission was to move up the north side of the valley and clear the small village of Warsak.[15]

To cover their initial move down the Ghaki Valley, Lieutenant Colonel Vowell's plan deployed a platoon from B Company, 1st Battalion, 327th Infantry, that had joined his battalion for the Afghanistan deployment. Along with their company tactical headquarters, Vowell deployed them on a five-hundred-meter hilltop overlooking the area surrounding the Marawara village, the start line for Stephen's platoon and the rest of Captain Weber's main force.[16]

Meanwhile, to block the Taliban's escape route to the south of Daridam, "the battalion Scout Platoon, led by Captain Kevin Mott, and reinforced with elements of the Omega force, would be placed by helicopter atop a high hill [LZ Hawk] overlooking Daridam." This position "was approximately 700 meters above the valley floor, 500 meters southwest of Daridam."[17]

For Stephen and his fellow soldiers, clearing the Ghaki Valley was going to be a tough fight against fierce resistance led by a formidable enemy commander. But unlike U.S. Army units in Kunar Province in

2008 and 2009 that had been outnumbered and outgunned, the Ghaki Valley assault force was well manned and well equipped with strong supporting air and ground assets.[18] Stephen, leading the spearhead of the attacking force, had fifteen soldiers in three military all-terrain vehicles (MATVs) and two MaxxPros as drivers and gunners. With bullet-proof glass and thick armor on the front, sides, and rear of the vehicles, the MATVs and MaxxPros were sturdy, resilient armored personnel carriers designed to absorb RPGs and IEDs and protect the soldiers within. The vehicles had M2 .50-caliber machine guns in an armored turret on the roof as well as Mark-19 Automatic grenade launchers.[19]

The MATVs also had a brand-new high-tech Common Remotely Operated Weapon Station (CROWS) that enabled the gunner to fire the vehicle's weapons system from the safety of the cabin. The difficulty was—as Stephen and the mounted elements of his platoon were soon to discover—that the smallest error in operating the CROWS caused it and the MATV itself to shut down. Stephen's platoon was light infantry, and his gunners had not had enough time to train on the CROWS system, leaving them at risk of being stranded amid enemy fire if the vehicle was disabled.

The remainder of the platoon was on foot. Stephen also had sixty Afghan National Army (ANA) soldiers as well as twenty Afghan Border Patrol troops. Immediate support came from the HHC's Capt. Steven Weber, the ground commander for the operation, with five MaxxPro MRAPs, the battalion's 120-mm mortar section, and the company's 60-mm mortar team. Alongside Captain Weber was the company fire support officer, 1st Lt. Spencer Probst. His task was to coordinate the air support and indirect fire missions: two Kiowas, two Apaches, and an Air Force sortie of F-15 Eagles as well as the battalion company mortars and a platoon of 155-mm howitzers from Camp Wright in Asadabad. Also in support were a platoon of Afghan "Tiger" commandos and the U.S. trained special counterterrorism force "Omega" and, for the end of the operation, a "clearing force," an elite company-size Afghan unit trained by U.S. special forces.[20]

The Battle for Daridam

Sunday, June 27, 2010, Ghaki Road, Ghaki Valley,
Kunar Province, Northeastern Afghanistan

A tree stood there, stately, old, its lower branches leaning over the Ghaki Road, an unpaved one lane roadway. It was the aiming point for Taliban insurgents. Once Stephen and his platoon reached it, intense Taliban fire opened up on them from an orchard down the road. It was the beginning of a twelve-hour firefight. By the end of Sunday, June 27, that ancient tree would be seared into the consciousness of Stephen and his platoon.

Forward Operating Base (FOB) Joyce

11 p.m. Saturday, June 26, 2010

At 11 p.m., precombat inspections began. After this, the U.S. and Afghan components of the Ghaki Valley assault force began to line up their armored vehicles in the order in which they were going to move out. This required integrating the ANA and the Afghan Border Patrol units into the U.S. column. This process was complicated by the shortage of interpreters and the separate U.S. and Afghan radio frequencies and because the Afghan forces had only been informed of the operation at 7:30 p.m. in order to ensure operational security. Simply put, the Afghan forces could not be fully trusted.[21]

In fact, operational security had already been compromised; the element of surprise had been lost. As early as 1 a.m. intelligence intercepts were picking up enemy radio chatter about the upcoming attack. Initial warnings had most likely come from villagers in the western part of the Ghaki Valley woken by the unusual volume of activity at FOB Joyce. Strong Eagle I was the only operation task force No Slack undertook in Afghanistan with a significantly large convoy component. And after the column moved out at 1:50 a.m. on Sunday, the noise of

the convoy's twenty-five vehicles and its sheer length—almost a kilo-
meter long—alerted villagers along the route, some of whom notified
the Taliban. Reflecting on this later, Stephen said, "There is no way they
didn't know. The Taliban is part of the fabric of society in the Kunar
Valley, and they watched FOB Joyce every minute of every day." It was a
pitch-dark moonless night, and the drivers—wearing night vision goggles and using the MRAPs' thermal vision cameras—drove through a
green-colored mountainous landscape.

At the same time that the column carrying the Ghaki Valley assault
force left FOB Joyce, the B/1–327 IN Company commander, Captain
Westfall, established a support-by-fire position about five hundred meters
above the Marawara village. His mission was to cover the initial movement of the assault force into the Ghaki Valley. On the ground below,
military police were posted to control access to the valley.[22]

A little after 3 a.m., the HHC section under Captain Weber left the
column. They established a command post at the partially completed
Afghan Police Station adjacent to the Marawara village. There, they also
set up the company's 120-millimeter mortar section to provide covering fire for the two infantry maneuver platoons.[23]

At the same time, the air assault operations began. In the inky blackness, Chinook helicopters touched down at three points in the Ghaki Valley. Alpha Company landed at LZ Owl on the northern ridgeline. Bravo
Company, reinforced by the company tactical command post, occupied
LZ Hen on the southern ridgeline. Both positions were between 1,000
and 1,500 meters above the valley floor. Their mission was to deny the
enemy the ability to emplace heavy machine guns to fire down onto the
two maneuver platoons on the valley floor. The third helicopter landing
dropped off the Battalion Scout platoon, reinforced by elements of the
Afghan Omega force under Captain Mott. They landed atop a ridgeline
700 meters above Daridam to prevent the Taliban from escaping.[24]

At about 4:40 a.m. the two maneuver platoons, the HHC, and the
Afghan forces occupied the Marawara village and waited for orders to
begin the clearance of the Ghaki Valley. They were joined by a U.S. Army
Engineer Route Clearance Platoon (RCP) with five vehicles including a

Husky, a Buffalo, and three MRAPs. The Husky was an armored mine/IED clearance vehicle that used sophisticated ground-penetrating radar in four panels at the front of the vehicle to detect mines. In appearance, it looked like a tan-colored elongated bulldozer.[25] The assault force was also joined by the U.S.-trained Afghan Tiger force with a mixture of Toyota's Hilux and Ford Ranger trucks.

Taking advantage of the pause, Stephen discussed final coordination arrangements with the ANA company commander. Lieutenant Jones did the same with the Afghan Border Patrol commander, whose troops were going to accompany 2nd platoon on the north side of the valley floor.

At 5:30 a.m., Stephen and the rest of the assault force had organized themselves in the order in which they would advance down the Ghaki Road. The ANA company on foot was in the lead, followed by Stephen and his platoon also on foot. Next came the Engineer RCP followed by Stephen's platoon. SFC Gabriel Monreal, Stephen's platoon sergeant, led the mounted section, with four MATV MRAPs, each equipped with two digitally automated .50-caliber CROWS machine guns and two MK-19 automatic grenade launchers. At the rear of the column came the Afghan Tiger force. The assault force for the north side of the valley floor also formed up with Lieutenant Jones's platoon, supported by the twenty-man detachment from the Afghan Border Patrol.

At 6:05 a.m. Capt. Weber gave the order to move forward. Lieutenant Jones's platoon began their advance toward the village of Warsak. On the Ghaki Road, Stephen's platoon began moving east toward the villages of Sangam and Daridam. The Taliban were waiting for them. At 6:46 a.m., Stephen learned over the company net that intelligence intercepts had established up to fifty insurgents in Daridam who were preparing to attack the two advancing U.S. platoons and their ANA and Afghan Border Patrol partner forces.

By this time, the ANA company arrived at the small village of Sangam. Stephen's platoon encountered no resistance as they helped the ANA company clear and secure the village. It quickly became clear that the population had fled. Despite tight operational security, news of Operation Strong Eagle I had leaked.

Just a few minutes after 7 a.m., the ANA company and the lead elements of Stephen's platoon passed the stately old tree that leaned over
the Ghaki Road. When they did, Rahman's Taliban forces opened fire
with RPGs, Russian-made PKM machine guns, and AK-47s.[26] The Taliban chose the place and the timing of the ambush with precision. They
had the perfect aiming point—the stately old tree—and the engagement zone was confined. Moreover, the Taliban opened fire at exactly
the point where the Ghaki Road narrowed and where there was minimal cover. On the left side of the road, as Stephen saw it, was a terraced
wall nearly ten feet deep ending in the wadi, or ravine. On the right was
a ten-to-fifteen-foot vertical natural rock wall. The Taliban timing was
perfect too. They did not try to stop the clearance of Sangam. Rather,
they held their fire until the ANA company and the lead elements of
Stephen's platoon were separated from the rest of the attacking force by
about two hundred meters, by a small hill and by a bend in the road.

The Taliban fire was intense and accurate. It began with two RPGs
and steady machine-gun fire from the Russian-made PKMs. The enemy
fire came from across the wadi, from *qalats* in Warsak, and from Daridam.
The RPGs killed one Afghan soldier and seriously wounded another
instantly, blowing him into the air and ripping away the lower part of
one leg. Staff Sergeant Shaw, leading Stephen's advance squad, immediately sprinted forward to help the ANA company commander respond
to the Taliban attack. The Taliban machine-gun fire claimed an American victim too. Pfc. Stephen Palu, Stephen's Bravo team M240 gunner,
wounded in his left thigh and right arm.[27] At the same time, on the
north side of the Ghaki Valley floor Lieutenant Jones's platoon had also
come under enemy fire, but it was less intense.

Reviewing the situation, Stephen immediately grasped the two elements of the problem. First, his advance squad accompanying the ANA
company were out of sight of the rest of the column, with its heavy
weapons that could provide effective suppressing fire. Second, there was
no cover at the ambush site and for about two hundred meters around
it. To his left was a sheer ten-foot drop into the wadi, and to his right
was a sheer vertical wall of rock. Lying in the prone position on the

direct road, two members of Staff Sergeant Shaw's squad—Pfc. Christopher Hendrix and Spc. Zachary Sturgess—returned fire against the insurgents in Warsack on the north side of the wadi with a M240B machine gun, but it was not enough.[28]

Stephen withdrew over the hill to reunite with the rest of the column and take advantage of the cover and the firepower that the MATVs could provide. He tried to report the situation and make his recommendation to Captain Weber over the company command net but could not get a response perhaps because of the mountainous terrain. With Taliban bullets kicking up dirt around his feet and taking chunks out of the natural stone wall, Stephen directed the counterfire and ordered his soldiers to withdraw in "buddy teams" to the limited cover of the armored personnel carriers.[29]

At this point, Stephen's platoon forward observer, Pfc. Dorian Clark, requested and received fire control of the two Kiowa scout helicopters to suppress the intense Taliban fire coming from Warsak and Daridam. Over the Company Fires Net, Private First Class Clark began identifying the grid squares so that Kiowas could target their suppressing fire as accurately as possible. This was difficult because neither Clark nor his platoon leader could spot the precise location of the Taliban firing positions. As a result, the air support was not as effective as it might have been, as the enemy utilized the *qalats* for cover and concealment.

At that moment, Staff Sergeant Shaw deployed his squad—Cpl. Adam Switchenberg, his M249 light machine gunner, his rifleman Spc. Jeremy Philipsack, and grenadier Pfc. Eric Soriano—just short of where the dead Afghan soldier lay on the road to protect his body and return fire. As they fired disciplined three-to-five-second bursts at the Taliban muzzle flashes, Shaw moved to help the ANA commander.

Staff Sergeant Shaw told his team that he would be right back. Under fierce accurate Taliban fire, he moved east down the Ghaki Road to help the ANA commander get his soldiers off the road into cover. After the Afghan soldiers began to move to positions on the terraces below the road or behind the rock on the southern ridgelines slope, Shaw then moved back to his squad's position. As the Taliban machine-gun fire and RPG rounds continued to hit the ground around Shaw and his soldiers, they began to move back toward armored personnel carriers in buddy teams. At the same time SSgt. Robert Livingston moved to Private Palu to provide his M240B gunner medical aide. "Using his own body to provide cover for Pfc. Palu, he began to treat his wounds as the AAF machine gun fire continued to kick up dirt and chip the stone wall beside them."[30]

Another example of selfless courage was Cpl. Joshua Frappier, who ran through ferocious enemy fire to help Staff Sergeant Livingston control the bleeding from Private Palu's wounds. Livingstone and Frappier then carried Palu through heavy enemy fire to Sergeant First Class Monreal's casualty collection point, behind the route clearance platoon.[31]

As the intense Taliban PKM machine-gun fire continued, Stephen ordered his M240B machine-gun crew, Specialist Sturgess and Private First Class Hendricks, to fire controlled six-to-nine-second bursts at the Taliban muzzle flashes. To provide additional suppressive fire and to allow the remainder of his soldiers to withdraw, Stephen told his observer, Private First Class Clark, to work with Sgt. Casey Cleveland, the fire support NCO with the HHC element, to get exact grid references for the Kiowas and Apaches. Cleveland also called in fire from the company's heavy mortar section at the ANP station and the 155-millimeter howitzer platoon at Camp Wright, fifteen miles away.[32]

Next to withdraw was Staff Sergeant Shaw, followed by the M240B SAW crew, Specialist Sturgess and Private First Class Hendricks. Only Private First Class Clark and Stephen were now left exposed to enemy fire. At about 7:25 a.m., Pfc. Clark withdrew, followed by Stephen.

At 7:30 a.m. Stephen's platoon had gained some measure of cover behind the Husky, the Buffalo, and the MATVs, but he had yet to report to Captain Weber. Because Stephen still could not reach the company commander by radio, he told his soldiers to remain under cover while he ran west down the road toward Sangam.

Stephen found Captain Weber standing between Sergeant First Class Monreal's MATV and the rock wall on the south side of Ghaki Road.

Stephen told Weber about the ambush and his intention to continue the mission in his platoon's armored personnel carriers accompanied by the Husky and the Buffalo of the Engineer RCP. His immediate goal was to help the ANA company by giving them cover and suppressing fire so they could treat their casualties and get some respite from the intense and accurate Taliban fire. Weber approved the plan, and Stephen ran back to his soldiers.[33]

A few minutes later Stephen led the advance on foot to the right of the first MATV. He was accompanied by Staff Sergeant Shaw and his forward observer, Private First Class Clark. Staff Sergeant Livingston followed, leading the M240B machine gunners carrying their four-foot-long weapons. SSgt. Joseph Moore's squad brought up the rear, advancing next to the third MATV.[34]

Staff Sergeant Shaw's Death

At 7:32 a.m., death came with cruel force to the 1st Platoon. Bursts of Taliban rifle and machine-gun fire had intensified, hitting the left and front of the lead MRAP with a loud metallic clang. Stephen positioned himself against the MRAP. A few feet away as Staff Sergeant Shaw began to turn toward the armored vehicle, he was shot through the left side of his jaw and neck. The force of the bullet twisted his body. He was dead before he hit the ground, but Stephen did not know this yet. Instinctively, he moved to help his friend and platoon mentor. He cradled Shaw in his arms, calling, "Eric, Eric."[35] Within a few seconds, however, Stephen saw that the Taliban bullet had ripped Shaw's jugular vein on the left side of his throat and that he had bled out. One of his soldiers nearby lost control of himself for a moment and began moving aggressively against the Afghan soldiers as if to harm them before Stephen stopped him.[36]

Losing a trusted friend in such a sudden, violent way was a profound shock for Stephen, his soldiers who witnessed it, and the rest of the platoon. Staff Sergeant Shaw was a beloved figure. But Stephen couldn't stop and mourn his friend. He was the leader of a platoon under intense enemy fire and could not show emotion. He suppressed the searing emotional pain and did his duty: giving his young soldiers experiencing combat for the first time the courageous leadership they needed. Stephen later told me, "What choice did I have at that point? I'm the leader. I can't stop. I wouldn't consider it courage. It's duty. It all comes back to knowing, living what right is and doing it. If you don't, more people will die." That is all true, but it had to be the hardest thing he had ever done in his young life.

There was continuous heavy enemy fire that kicked up dirt and snapped through the air all around Stephen. Bursts of machine-gun fire flew by in clusters. Despite this fusillade, Stephen lovingly lifted up Staff Sergeant Shaw's lifeless body by his armor's shoulder straps and brought him behind the MRAP. Spc. Jeremy Phlipsak and Stephen removed Shaw's equipment and gently lifted his body into the back of the MRAP. Stephen reported the death by battle roster number to Sergeant First Class Monreal, who asked, "Who?" Instead of giving the roster number again, Stephen replied "Shaw" over the radio. It was a gut-wrenching moment.[37] Furthermore, Staff Sergeant Livingston approached Stephen, who's uniform was covered in blood, and began to triage him to ensure that he was not wounded. He wasn't, but Livingston performed his duty to perfection to confirm that his platoon leader did not require aid.

Around 8 a.m. Stephen, accompanied by Staff Sergeant Moore, Corporal Frappier, Specialists Howard and Sturgess, and Private First Class Clark, put Staff Sergeant's Shaw's body on a backboard, covered it so that other soldiers in the platoon would not see his disfigured face, and brought him to the casualty collection point behind the RCP's armored vehicles. There, they transferred him to a Toyota Hilux truck, sent to bring the dead and wounded back to the HHC casualty collection point at the Marawara village. A medevac helicopter could land safely there and evacuate Shaw's body and the wounded Private Palu who had been sent back earlier. Moore volunteered to accompany Shaw's body and saw it safely loaded on board the medevac helicopter and unloaded upon landing at the Marawara village.[38]

Back in the kill zone, a twenty-minute smoke barrage from the 155-millimeter howitzers at Camp Wright gave Stephen's platoon some concealment and temporarily helped reduce the rate and accuracy of enemy fire. Once the smoke dissipated, however, Taliban fire resumed and was as relentless and effective as before. Rahman's forces were firing RPGs and PKM heavy machine guns with great accuracy. An RPG hit the engine block of the RPC's Buffalo, disabling it. The Taliban had skillfully targeted the Buffalo's most vulnerable point. The Ghaki Road was now blocked.[39]

For Stephen's platoon, there was one advantage, however. They now had the combined firepower of the RCPs and three MRAPs along with those of his own platoon's MATVs. Stephen's own gunner, Pfc. Thomas Shelton, identified a Taliban RPG team about seven hundred meters away. With two six-round bursts from his .50-caliber system, he killed two Taliban insurgents and disabled their weapon. Another gunner, Pfc. Thomas Wade, fired single rounds manually from his .50-caliber machine gun. It was all he could do because the gun kept malfunctioning in the heat, and there was no opportunity to gather additional lubricant or conduct hasty maintenance. As enemy bullets whizzed by his head and upper body, Wade tried to fix the malfunctioning machine gun.[40]

At the same time, around 10 a.m., the two Apaches assigned to the assault force arrived on station. Using their heavy 30-mm chain guns, the two Apaches began to strafe the pomegranate orchard north of the Ghaki Road from which such heavy Taliban fire had been coming. These attacks continued for two hours.

As the Apaches began to strafe the Taliban in the pomegranate orchard, Stephen jumped down into the wadi where wounded ANA soldiers were lying. Private First Class Meyers and Private First Class Hendricks triaged them and along with Stephen helped them up to the column and on to the casualty collection point.

Now Stephen turned his mind to the problem of the disabled Buffalo. As Taliban machine-gun fire on his position continued, he and Staff Sergeant Livingstone began to think through how best they could recover the disabled vehicle. Stephen also reported the problem to Captain Weber.

At this time, the Taliban counterattacked. They not only attacked all of the support-by-fire and overwatch positions on the valley's northern and southern ridgelines but also intensified their fire on Lieutenant Jones's platoon on the north side of the wadi. Clearing the Ghaki Road quickly and pressing on to Daridam was now urgent to ease the pressure on the U.S. troops on the ridgelines. The battalion TOC ordered Captain Weber to tell Stephen to push the disabled Buffalo off the road into the wadi. When he received the order, he thought it was madness. Destroying a multi–hundred-thousand-dollar vehicle made no sense to him. He told Weber that despite the enemy fire, he and his soldiers could attach a tow bar and cables to the disabled Buffalo in the same time it would take to strip the vehicle of its sensitive equipment and push it into the wadi. Weber approved Stephen's plan.[41]

A little after 11 a.m. under Stephen's leadership, his soldiers successfully removed the disabled Buffalo from the kill zone. In the village of Sangam, the RCP crew got the Buffalo turned around and towed it out of the Ghaki Valley. The courage, determination, and focus of the 1st Platoon was not mirrored in their Afghan partner force, however. Demoralized and shocked by their casualties, the ANA company refused to fight anymore and left the battle zone. The Afghan Border Patrol unit, however, stayed with Lieutenant Jones.[42]

For Stephen, the departure of the ANA company meant that his combat power had fallen from seventy-four soldiers to thirteen. Undaunted, Stephen consolidated and reorganized his platoon. Staff Sergeant Livingston now led those on foot. Stephen took charge of the four MATVs. Just before noon, he led his platoon through the junction of the wadi and the Ghaki Road.[43] The temperature had now risen to over 100 degrees Fahrenheit. At the HHC location, Captain Weber had had to have intravenous injections of saline to cope with dehydration. Under constant intense enemy fire, Stephen and his soldiers had no such opportunity. They had to keep fighting with the limited water supplies they carried.

As Stephen led his platoon down the Ghaki Road, the Taliban attacked his MATV with RPGs and PKM machine guns from the southern ridgeline and from a pomegranate orchard just ahead of his column on the north side of the wadi. He paused to allow his gunner, Private First Class Shelton, to fix the CROWS system, which had been malfunctioning. Once the system was repaired, Stephen moved east with his four MATVs to establish support-by-fire positions for his own foot soldiers and those of Lieutenant Jones's platoon. Pressing on to Daridam as soon as possible had now become important, as it was the ultimate objective and also was needed to relieve the pressure on Captain Mott's scout platoon, which was under heavy Taliban attack on the southern ridgeline.[44]

Within minutes, Stephen and his soldiers had entered a new and more dangerous landscape. The Ghaki Road crossed the wadi from south to north, and within two hundred meters of that junction lay the first of a series of *qalats* that marked the western border of Daridam. *Qalats* were fortified housing compounds built atop terraces full of pomegranate trees. Made of brownish-yellow adobe bricks, the *qalats* of the Ghaki Valley—like those elsewhere in eastern Afghanistan— provided ideal cover for the Taliban. The *qalats* of Daridam were camouflaged by the lush foliage of the densely packed pomegranate trees now in full bloom. As they drove slowly into the western outskirts of Daridam, neither Stephen, his MATV driver Private Hanawalt, nor his gunner Private First Class Shelton could identify where the insurgent positions might be.[45]

A few minutes before 1 p.m., disaster struck. As Stephen later recalled, "After moving at an idle along the first twenty-five meters of the pomegranate orchard to our left, our MATV suddenly shook violently; the engine stopped, and smoke began to pour in through the air vents. None of us could see the AAFs firing points through the narrow MATV windows. At 12:52 I reported to Capt. Weber that my MATV and CROWS system was disabled by RPG and machine gun fire."[46]

Stephen was trapped. The Taliban continued to pour RPG, machine-gun, and rifle fire at his disabled MATV. His CROWS system malfunctioned and then stopped firing altogether. At 1 p.m., Stephen told Captain Weber that he needed suppressing fire on the Taliban.[47]

Immediate air support came from two Kiowas. Over his platoon net, Stephen directed the pilots to where the Taliban were firing from. He requested that high-explosive rockets be fired into the *qalats* and that flechette rockets be fired into the pomegranate trees. Stephen hoped that the flechette rockets—a lethal antipersonnel weapon each one of which released up to ninety-six little arrows about the size of a dart before hitting the ground—would decimate the Taliban insurgents. As the Kiowas prepared to attack, the Taliban fired intensely and accurately at Stephen's already crippled MATV. Sustained PKM machine-gun fire hit the left, front, and right sides of his vehicle with a shrill clang, causing the thick bullet-proof glass to begin to crack. Worse, Taliban RPGs kept hitting the engine block, Worse still, hand grenades exploded on the roof and hood. Inside the cabin, Stephen and his two soldiers were protected, but the noise was deafening, and the threat was ominous. The Ghaki Road was now blocked for the second time that day.[48]

At 1:30 p.m. the Kiowas began their attack runs against the Taliban insurgents in the *qalats* and pomegranate orchards. They broke station only after Stephen's battalion command, Sgt. Maj. Christopher Fields, arrived to provide additional suppressive fire with four MaxxPro MRAPs. Ten feet tall, heavily armored, IED-resistant, and equipped with one heavy .50-caliber machine gun mounted in a turret, the MaxxPro MRAP had an imposing presence. In all, Sergeant Major Fields brought four heavy machine guns to provide extra suppressive fire so that Stephen and his soldiers could recover his crippled MATV and get it out of the battle zone.

Despite continuing Taliban fire, one round of which gave Sergeant Major Fields a flesh wound in his calf, he and Stephen attached a tow strap from the lead MaxxPro to the disabled MATV. Their first attempt failed because the wheels of the MATV locked repeatedly. The problem was that the drive shaft was locked, and so further attempts to recover Stephen's disabled MATV also failed. There was only one thing left to do: strip the crippled MATV of sensitive equipment and abandon it.

Under fierce Taliban fire, Stephen and Sergeant Major Fields, joined by Private First Class Hanawalt and Private First Class Shelton made three trips between the lead MaxxPro and the crippled MATV to remove the sensitive equipment. On the fourth and final trip between the two armored vehicles, they locked the gunner's hatch and all the doors to prevent the Taliban from getting into the MATV and taking any remaining equipment or ammunition.

But as they began to withdraw, disaster struck again: Taliban RPGs and PKM heavy machine-gun fire crippled a second of the 1st Platoon's MATVs. Luckily, the two Kiowas came back on station. Refueled and rearmed and guided by gunner Shelton, they provided suppressing fire so that Stephen and Sergeant Major Fields could try to recover the second MATV.[49]

Supported by Private First Class Hanawalt, Stephen and Sergeant Major Fields moved to the rear of the second disabled MATV, approximately twenty-five meters west of the first, supported by the MaxxPros. Describing their actions under heavy enemy fire, Stephen later wrote,

> We attached a tow strap from the front of the MAXPRO to the hitch on the MATV. PFC Wade, the gunner for the second disabled MATV, continued to engage the pomegranate orchard with his .50-caliber machine in the turret. The insurgents' machine-gun fire continued to impact the left and front sides of the vehicle while we attempted to drag it westwards. The engine of the second MATV continued to run; however, the power steering unit had been disabled, so the vehicle began to slide towards a drop-off and down a terrace on the right side of the road as it slid backwards.[50]

They stopped the second disabled MATV from sliding off the road and replaced the tow strap with a tow bar that gave them greater control.

At this point, Sergeant Major Fields decided to try to pull both MATVs out together, so he and Stephen attached the tow strap between the two disabled MATVs. Stephen later recalled that "throughout the recovery process, we continued to receive small arms fire from the insurgents occupying Daridam—the rounds impacted the vehicles around us and ricocheted underneath them past our boots. Once the tow bar and tow strap were attached between the three vehicles, CSM Fields told me that he needed me to re-enter the first MATV to keep the wheels straight, even though they would rotate, to help steer it out of the engagement area."[51]

Stephen's driver, Private First Class Hanawalt, insisted that it was his job to steer the crippled vehicle. But Stephen would not allow one of his soldiers to run through a hail of enemy fire alone. They would run to the disabled MATV together, open it, climb in, and try to steer it out together. At first they made progress, but because the engine block was so badly damaged by RPGs, the wheels remained locked.

By this point, Stephen and Sergeant Major Fields faced a serious problem: the two MATVs were so badly damaged that they could not be recovered without external help. At the same time, it was vital to extract the remaining undamaged MATVs and MaxxPros from the immediate battle zone lest they too be disabled by Taliban RPG and heavy machine-gun fire. The only realistic choice was to abandon the two crippled MATVs and withdraw from the battle zone.[52]

By 5:15 p.m., Stephen had consolidated all remaining members of his platoon at the junction of the Ghaki Road and the wadi. Two hours later as the sun began to set, Lieutenant Jones's platoon joined them. The Taliban, meanwhile, kept up sporadic bursts of fire from Daridam. With the arrival of darkness, the 1st and 2nd Platoons got their first reprieve from constant Taliban fire in twelve hours as well as the from the extreme heat of the past six hours. But this welcome respite was tempered by the disheartening news of the withdrawal of the Afghan Tiger and Omega forces from the Ghaki Valley. It was yet another sign of the unreliability of the Afghan forces.

By nightfall, the two young lieutenants had to face facts: between their two platoons, they could muster fewer than forty soldiers, not enough to clear the Ghaki Valley and take Daridam. They needed more troops and heavy air support to dislodge the Taliban from Daridam. The extra troops came in the chilly early-morning hours of June 28: an

Afghan commando company accompanied by their special forces U.S. military advisers arrived by helicopter at a hastily cleared landing zone near the Marawar village. The heavy air support also arrived after midnight: a formidably armed AC 130H Spectre gunship. Aptly nicknamed the "Angel of Death," the gunship pulverized Daridam throughout the night with its four 30-millimeter cannons, its 105-millimeter howitzer, and its 20-millimeter Gatling gun, killing large numbers of Taliban insurgents.[53]

At about 3:30 a.m., Captain Weber briefed Stephen, Lieutenant Jones, and the Afghan platoon commanders on the plan to attack and clear Daridam shortly after sunrise. The plan was that the Afghan commandos would come in toward Daridam along the northern ridgeline and establish a support-by-fire position overlooking the village. Meanwhile, the 1st and 2nd Platoons would approach Daridam through the wadi from the south. At 4 a.m. Stephen led his platoon down into the wadi so they could use the remaining darkness to mask their movement toward the Taliban stronghold. Once they found and occupied a secure assault position in the wadi, they settled and waited for sunrise.[54]

When it came, Stephen's platoon was surprised to discover that the Taliban had abandoned Daridam. Apparently, they had used the temporary absence of U.S. air cover after the departure of the Spectre gunship to escape farther east up the Ghaki Valley toward the Pakistani border. Once they entered Daridam, Stephen and his platoon found the village deserted: no people and no animals, only the corpses of the Taliban and livestock killed during the previous day's twelve-hour firefight and the overnight bombardment by the Spectre gunship.[55]

For Stephen and his platoon, this was victory. It demonstrated Stephen's courage, commitment, and inspirational leadership. Only his four NCOs had experienced combat before. Stephen's soldiers were young with no combat experience at all, yet they had performed courageously and effectively. This validated not only Stephen's leadership but also the quality and effectiveness of the platoon's rigorous training, discipline, and unit morale.

The death of SSgt. Eric Shaw, however, was a terrible blow. He had died selflessly trying to save Afghan soldiers and protect his own. In the years following, the Department of Defense would award him the Distinguished Service Cross for his actions to save not only American lives but also those of our Afghan counterparts. Shaw's death was emotionally devastating for Stephen, but he couldn't stop and mourn his friend. He was the leader of a platoon under intense enemy fire and could not show emotion. Drawing on his training and exceptional moral character, Stephen suppressed the searing emotional pain and gave his young soldiers the courageous leadership they needed to survive and complete the mission during the weeks that followed.

PART 5
———
EPILOGUE

24

NICK ESLINGER

On Thanksgiving Day 2008, Nick and his soldiers returned to Fort Campbell, Kentucky. After completing his postdeployment duties, he took three weeks of postdeployment leave in Houston with his parents, Donna and Bruce. In early January 2009, Nick returned to the 1st Platoon at Fort Campbell and again undertook the training cycle, beginning on the marksmanship ranges. It was a time of transition within the unit. A number of leaders processed out to different duty stations. New leaders and a crop of new soldiers arrived, so Nick's immediate priority was to train them up.

At the end of January, Lt. Col. Joseph McGee called Nick into his office and asked him if he had ever considered applying to the Army's elite Ranger Regiment to be a platoon leader there. Nick replied that he had thought about it and would love to. Lieutenant Colonel McGee said he would support it, adding, "Why don't you fill out an application and get it to me in a week?" With McGee's strong support and his own impressive record, Nick was selected for the Ranger Orientation Program. This program is two weeks of physical and mental stress designed to sort out which volunteers have what it takes to join the Army's 75th Ranger Regiment, an elite special operations unit that conducts covert raids in enemy territory around the world. In April 2009, Nick reported to the Ranger Orientation Program shortly after he had received the Silver Star from Gen. George Casey, chief of staff of the Army.

In June 2009, having successfully completed the Ranger Orientation Program, Nick reported to the 2nd Ranger Battalion at Fort Lewis, where he remained for two years with the Rangers. In December 2009, he deployed to Afghanistan as a Mortar Platoon leader. His job was to serve as liaison officer to the 2nd Marine Expeditionary Brigade at Camp Leatherneck in Helmand Province. Sections and teams from Nick's mortar platoon were spread across Afghanistan supporting Ranger elements in all the various regional commands in the country. The operational details of the Ranger Regiment are classified and cannot be reported here.

Nick redeployed to Fort Lewis in March 2010 and remained there until he took the next step for any young army officer, the Maneuver Captain's Career Course at Fort Benning, to prepare for company command. Nick found the six-month course invaluable because it gave him the opportunity to deepen his professional knowledge of Army doctrine, military history, and tactics. Above all, Nick thought that he developed a much deeper knowledge of tactics.

Newly minted as a captain, Nick moved to Fort Hood, Texas, with his new wife, Calisse, where he briefly served as a staff officer at 3rd Cavalry headquarters before taking command of the "Crazy Horse Troop." After nineteen months in command, he departed for San Diego to go to graduate school.

In 2013, Nick had applied to return to West Point as an instructor in the military leadership department. He wanted to give back to his beloved alma mater and saw teaching military leadership there as a good way to do it. The USMA granted his application but told him that he would need a master's degree in industrial organizational psychology before he could join the USMA faculty. To get his degree, Nick would attend San Diego State University.

The degree was highly relevant to not only his post at West Point but also his growth as an army officer. The industrial component of the degree was designed to teach the statistical side of organizational leadership by focusing on human behavior: quantitative analysis of everything from resilience to self-efficacy. The organizational component of the degree focused on understanding and assessing human behavior in the workplace. Both components of the degree were directly relevant to being an army officer. Nick also wrote a thesis on promotions outcomes for NCOs, that is, how these administrative choices affected the later active-duty performance of those promoted.

In going to a civilian graduate school, Nick was leaving the military bubble in which he had lived ever since attending the USMAPS in 2002. It was a big step outside his comfort zone because he was also walking into a totally different cultural environment in which none of the students in his master's classes shared the norms and values of the military culture he lived by.

Essentially, what Nick thought important, his classmates did not. As a young officer, he attached great importance to being in the right place at the right time and with the right materials. His classmates did not. Not using a smartphone when the professor was speaking was important to Nick: for him it was a matter of showing respect. Civilians used smartphones in class and thought nothing of it, and Nick soon found out that in carrying out group projects, civilians did not believe that everyone should put in equal effort. All too often, he would find himself doing most of the work while many of his fellow graduate students were bystanders. When group papers were assigned, he expected his classmates to join a meeting where they would talk through and agree what each member of the group would contribute to the project. This is what happened in the Army. It was not what happened at San Diego State. There, civilian graduate students preferred to be isolated at home and make their contributions online. For Nick, controlling his reaction to these different cultural norms was sometimes difficult.

The civilian graduate students had different perceptions, different perspectives, and different approaches to solving problems. For Nick, this was also a big intellectual challenge: he had not been in a university class since graduating from West Point in May 2007. In contrast, most of his civilian classmates had just received their bachelor's degrees and were academically sharp. In terms of academic learning Nick was a little rusty, and he knew it. As he later recalled, going into a graduate-level statistics class with recent math graduates was humbling.

Overall, Nick found his time at San Diego State a great developmental experience. It not only sharpened his intellect but also gave him an understanding of a whole new field of knowledge. His graduate school experience gave him that and so much more: greater self-awareness, the opportunity to assess his strengths and weaknesses in relation to a different peer group, and different ways to think about problems in human and organizational behavior. In short, his San Diego State experience gave him a greater openness of mind and a greater tolerance in dealing with civilians, qualities that would serve him well in his future military career.

At West Point, Nick taught as an instructor in the Department of Behavioral Sciences and Leadership. He felt honored and proud to have the opportunity to give to his alma mater, which had given him so much. As he recalled, "It took me from being an eighteen-year-old with no clear path to follow and provided an opportunity which I seized. It helped me become the man I am today and provided a future for the family I now have." For two years, Nick helped shape the way cadets thought about leadership.

In 2018 Nick continued his own professional education at the U.S. Army Command and General Staff College at Fort Leavenworth.

On May 7, 2019, Maj. Gen. Stephen Townsend, commander of the U.S. Army's Training and Doctrine Command at Fort Leavenworth, presented Nick with the Distinguished Service Cross. This is the Army's second-highest award for combat valor in which a soldier puts his own life at risk to save others. The award was the result of a Pentagon-wide review of awards for valor in the war on terror launched by Defense Secretary Jim Mattis. The goal was to see which awards for valor already made needed to be upgraded. Nick's Silver Star was one of just over a dozen awards chosen for upgrade.

In his remarks at the ceremony, Major General Townsend said, "I think you demonstrated by your actions that day that you were willing to die for them if necessary, to sacrifice all of your tomorrows for your soldiers. You are an inspiration to all of us, and I am honored to serve in the Army that produces such leaders."[1]

Late in 2019 I asked Nick whether the prolonged combat stress and the death and destruction he had witnessed had taken a personal toll. He said, "No, not in any significant way. I never had nightmares, much less PTSD." Nick explained that when he deployed, he wanted to stay there. While not diminishing the horror of war or comparing war to sport, he used a sporting analogy to explain his thinking:

Imagine a team of football players who dedicate themselves every day in practice to be the best they can possibly be but never gets to play an actual game against an opponent. In garrison, we train and we train to be the best we can possibly be. When we finally deploy on the battlefield, we're finally doing what we were trained to do. We're doing our job. We're serving and protecting the United States against our adversaries. And so, there was always a part of me that did not want to go back to practice. Deploying is what I signed up for: to serve wherever the nation needs me to serve, and to lead there.

25

TONY FUSCELLARO

In April 2010, Tony Fuscellaro's squadron returned to Fort Bragg. After a year of intense fighting in Afghanistan, Tony was now safe and able to live together with his bride, Adrienne, for the first time. He continued to serve as Lt. Col. Mike Morgan's adjutant until departing Fort Bragg to take the Aviation Captain's Career Course at Fort Rucker, Alabama. After completing the course, Tony applied for the elite 160th Special Forces Air Regiment. He was accepted to fly Chinooks but declined because he thought flying such heavy transport aircraft did not fit his personality, much less his experience as a ground-attack pilot.

Instead, Tony went to Savannah, Georgia, and took command of Charlie Troop, 3rd Squadron, 17th Cavalry, and in December 2012 he led them back to Kandahar, Afghanistan. There, he went back to live in the same plywood buildings at the side of the Kandahar airfield that he had helped build three years earlier and to fight in the same area.

When Tony deployed back to Afghanistan, he did so with what he described as "an immense level of comfort." He felt this comfort not only because of his previous combat experience and because he had studied the landscape of Kandahar so carefully but also because he knew how the Taliban fought and knew his Kiowa helicopter really well.

In his second deployment, Tony brought the supreme professionalism he had been taught by Lieutenant Colonel Morgan: not only the
training to reach near-perfect firing accuracy but also the painstaking
preparation getting to know his area of operations, the same meticulous after-action analysis we saw earlier. By 2012, however, the Kiowas
had an additional advantage: every FOB and every unit had minicam-
eras mounted on balloons that could go up to five hundred feet.

The Kiowa troop that Tony now commanded was part of the 3rd
Squadron, 17th Cavalry. The squadron as a whole had thirty Kiowas:
three troops with ten aircraft each, a maintenance troop, and a forward
support troop for refueling and rearming. Tony's troop, C Company,
had ten Kiowas and sixty soldiers. They too had combat experience but
it was in northern Afghanistan where the high mountains and the bad
weather made combat flying different. Simply put, combat flying amid
high mountain peaks with severe weather was different than flying over
the desert or the lush Arghandab and Helmand River valleys. In the
mountains, you fought at a thousand feet. In the desert and river valleys, you fought at a hundred feet. In Kandahar, a Kiowa could be easily
spotted and shot down at a thousand feet. So, Tony had to teach his
Kiowa pilots a new way of fighting.

In Kandahar between 2010 and 2012 the intensity of the fighting
had decreased significantly, and the location of the fighting had changed.
Tony led the brigade in a number of engagements with the enemy, but
he had only about a dozen serious fights with the Taliban throughout his deployment. That compared to 2009–2010, when a dozen fights
would have been a quiet month. The location too had changed. In
2009–2010 most of the fighting took place in the Arghandab Valley
and south of Highway 1. In 2012–2013, Tony had a small number of
engagements in the Arghandab Valley and south of Highway 1, but not
a single fight came remotely close to those of 2009–2010 in terms of
scale and intensity. By 2012–2013, the fighting had moved south.

Tony did not complete his yearlong deployment because in the
summer of 2013, he was redeployed out of command to take up a fellowship at the Strategic Studies Institute at the University of North

Carolina and Duke University. The program—previously reserved for one-star generals—gave Tony his first real exposure to the strategic level of war and to the study of civil-military relations and strategic studies more widely. Tony thought that the program was "phenomenal." He loved to study and found that that program not only deepened his knowledge but also enabled him to think critically and creatively beyond the tactical level.

During his time in North Carolina, Tony learned about, applied for, and was accepted to the Joint Chiefs of Staff fellowship. This is a highly competitive program that gives outstanding young army captains three years of intellectual and professional development at the Georgetown University Public Policy Institute, the Office of the Secretary of Defense, and the Army staff or the Office of the Secretary of the Army. Tony's program did not begin until January 2014, so after completing the Strategic Studies Institute program he went back to Savannah as an assistant brigade S3, where he did operational brigade-level planning.

In December 2013 Tony and Adrienne moved to the Virginia suburbs outside of Washington, D.C., to begin his fellowship. It began with a shortened version of the Army's Command and General Staff College curriculum, which he took at the Staff College's satellite campus at Fort Belvoir. Shortened from two academic semesters to one, the program was designed to give young officers the opportunity to complete their education at a civilian university. For Tony, as a Joint Chiefs fellow, that meant Georgetown University, where he studied for a master's degree in policy management.

Tony loved Georgetown, where, as he put it, he "learned a ton." He was a far better student at Georgetown than he ever was at West Point. The intensive study required during his operational deployments were so demanding that he found Georgetown "a breeze." Tony took multiple classes in foreign policy and in American government. He learned about the inner workings of the U.S. government: budget cycles and how the White House, the rest of the executive branch, and Congress work together. The biggest lesson he drew from his Georgetown studies was that if military officers want to have influence at the highest level of government, they need to have different tools in their kit bag. Typically, the skills and attributes that help you become a successful military officer—caring for your soldiers, motivating them, and leading them with courage and willpower—are not the skills you need in the academic or policy-making world. If anything, these military characteristics turn some people off. So, military officers have to learn to be more detached, more measured in the way they approach complex public policy problems.

Another lesson Tony drew was the importance of understanding the culture of young civilian millennials: how they thought, acted, and functioned. Those he met at Georgetown were exceptionally bright young people fresh from their bachelor's degrees. They had no real work experience, but soon after graduating with their master's degrees they would be working for the Department of Defense, the Department of State, or the intelligence community and would begin influencing policy. For Tony, it was therefore very important to understand these young millennials: how to discuss and debate foreign and national security policy with them and find common ground and compromise.

The next phase of Tony's fellowship brought him to the Office of the Secretary of Defense, headed by Ash Carter. There, Tony worked in the program analysis division of the Office of Cost Assessment and Program Evaluation (CAPE). Put simply, CAPE's mission is cost control and budget savings. CAPE's job was to study all of the programs of the Department of Defense, and, as Tony put it, "peel them apart within the budget cycle, check that they were receiving appropriate amounts of funding and spending the appropriate amount." Perhaps even more important, Tony and his colleagues also worked to ensure that these programs were properly resourced over the next five years.

Within CAPE, Tony worked in the land forces division, which controlled the Army and focused on aviation programs. Specifically, he worked on the replacement for the aging Kiowa warrior, the upgrades to the Black Hawk and the Apache, a new improved turbine engine for army helicopters, and the Integrated Terrain Analysis Program, a new improved visual sighting system that will enable pilots to see through heavy clouds and dust storms. In each case, Tony's task was to determine future funding needs and progress through the Army's acquisition system.

For Tony, a combat flyer with two tours in Afghanistan behind him, this job was ideal. In a meeting when his colleagues said "let's discuss the Integrated Terrain Analysis Program," for example, it was important for them to hear Tony speak about the vital importance of developing this new system from a combat pilot's perspective. In Afghanistan, he had to fight with a Kiowa that had no electronic sight through dense cloud cover, much less sandstorms. When the CAPE examined the importance of a new improved turbine engine for army helicopters, Tony again brought the combat flier's perspective, having flown the agile but underpowered Kiowa in Afghanistan. He could explain to his colleagues how vital it was to have a more powerful engine that could provide new capability in mountainous high-altitude environments as well as over deserts and densely forested landscapes.

For the third year of his Joint Chiefs of Staff fellowship, Tony moved to the Office of the Secretary of the Army. There, he worked directly for Army secretary Eric Fanning in his front office, preparing Fanning for meetings with members of Congress as well as senior Army officers. In addition to briefing the secretary, Tony's duties included preparing follow-up notes and actions. This was a great learning experience for him because it was his first exposure to a senior civilian executive, a political appointee who had spent nearly fifteen years in the Department of Defense and had excellent connections throughout the Pentagon and with President Barack Obama and his White House team.

Tony also enjoyed learning how a highly disciplined executive such as Secretary Fanning ran his office. Years later Tony recalled, "I will never forget telling four stars that they were looking at seven to ten minutes for the meeting. They would smile at me, a lowly major, saying 'I'm going to be in here an hour. We have a lot to discuss.' Secretary Fanning would stand up and walk away from the table after seven minutes."

Tony learned so much by observing how thoroughly Secretary Fanning prepared for meetings days or weeks before. And when

Fanning finally took a meeting, he had very precise questions he wanted answered. Once he had his answers, there was nothing further to discuss. "This helped round me out," Tony said years later.

Late in 2017, after a year with Secretary Fanning, Tony transitioned back to the field army. He returned to where he had begun his military aviation career: the 1st Squadron, 17th Cavalry, at Fort Bragg. The 1st Squadron was no longer a Kiowa squadron, and Tony arrived as they were beginning to "stand up" as an Apache unit. He helped the squadron receive their first Apaches, integrate them, and train up new Apache pilots. For seven months he became squadron S3, the operations officer, before moving to squadron XO for another six to seven months. Early in 2019, Tony became the XO of the 82nd Airborne's entire aviation brigade. Soon he was chosen for early promotion to lieutenant colonel, the first of the elite souls to earn that rank.

In May 2019 I sat with Tony in his home near Fort Bragg as he reflected on his career to date. On that sunny May morning, I asked Tony whether the prolonged combat stress and the death and destruction he had witnessed had taken a personal toll. He said it had but that he had never had nightmares, much less PTSD. Our discussion revealed that the main reason he did not suffer from nightmares or PTSD was because his actions as a combat pilot reflected his own moral values. He said, "I was very confident that I was always doing the right thing; I never pulled the trigger unless I was sure. We were very calculating in the way we engaged and when we got into a fight. So, I don't have a single engagement that I go back on and say that one was close or ambiguous, maybe I should not have shot here. It doesn't exist." He added, "Morally, everything I shot at was in defense of myself or someone else. We were fighting *very* evil people. There were so many horrible things that I saw, but I put them to the back burner, because I knew that our presence out there saved lives. Every time we were in a fight it was because we needed to be. Among the Taliban there was a shocking level of moral evil."

This was the impetus behind Lieutenant Colonel Morgan's decision to engage the Taliban during their attempted ambush of his Kiowa squadron on Christmas Eve, 2009. He realized that some of the Taliban's most evil commanders were on the ground in Howz-e-Madad. Without them, the complex triangular ambush would not have occurred. They had to be eliminated.

I believe there was a second reason Tony did not suffer nightmares and PTSD: the way his CO on his first deployment to Afghanistan, Lieutenant Colonel Morgan, framed their mission as a moral issue. Their goal, he told his pilots repeatedly, was to save soldiers' lives. With deep respect for the concept of just war, Morgan repeatedly emphasized respect for the dignity of human beings coupled with respect for the rule of law. This is another reason why Tony believed that he could carry out his mission with moral comfort.

In the spring of 2019 Lieutenant Colonel Morgan, Tony's mentor and role model, retired from the army after a long and distinguished military career. At about the same time, Tony received early promotion to lieutenant colonel. His next assignment will be as deputy commander of the 82nd Airborne's Air Brigade. To produce an outstanding battalion commander of Morgan's stature takes up to twenty years of education, experience, and training. It takes an enormous amount of development to build tactical competence, learn how to build a true moral compass, and learn how to take care of soldiers and become an effective commander who can lead a unit to the highest standards. Over a decade ago, Morgan inspired and trained Tony, then a lieutenant. Now, after two years of intensive combat experience rounding out at the strategic and operational level, Tony is ready to step into his honored mentor's shoes: developing young lieutenants and captains and inspiring and leading a unit to its highest level of effectiveness.

26

ROSS PIXLER

On May 16, 2008, Ross Pixler's platoon as well as the rest of the battalion returned home to Fort Benning. Ross described the welcome home ceremony as "most special." After turning in their weapons they marched into the terminal, where they were congratulated by senior military officers and civic leaders. After the ceremony, Ross and his soldiers were released to their families. Ross's wife, April, and their eighteen-month-old daughter, Dakota, were there to greet him. April cried tears of joy and gave Ross an enormous hug. Dakota, however, started screaming and crying. She had been only three months old when Ross left for Iraq and did not recognize her father. Worse, she had never seen a man touch her mother before and thought this strange soldier was trying to hurt her. It took a while for Dakota to accept Ross as her father.

Ross remained with his platoon from May until July 2008, when he was promoted to captain and transferred to Baker Company, Hardrock's sister company, as XO. SFC Pete Black was promoted and transferred with him as company first sergeant, so they were able to continue to work together as a team. Unlike Hardrock, Baker Company had seen very little action in Iraq and usually arrived in a sector that Hardrock had just cleared.

In December 2008 Ross tried out for Special Forces Assessment and Selection Course but failed on a technicality. In Iraq, an Army doctor had prescribed a medication to help improve Ross's breathing, which had been affected by smoke from large garbage-burning fires. Several days into Special Forces Assessment and Selection Course a member of the team evaluating candidates took Ross aside and asked him if he was still on this prescription. Ross said, "No. You told me I couldn't take it, so I stopped taking it. It's locked up in my civilian bag." Despite this, the evaluator told him he couldn't continue but acknowledged that it was a silly bureaucratic issue. The evaluator advised Ross to just wait until the next class and not put that prescription down on his paperwork. Ross refused. He was not going to lie just to get in.

Ross returned to his XO position at Baker Company until March 2009, when he attended his Maneuver Captain's Career Course at Fort Benning. In September 2009, he graduated and joined the 10th Mountain Division in Fort Drum, New York.

In March 2010, Ross deployed with elements of the 10th Mountain Division in the surge to Afghanistan. He spent ten months in a senior staff position. He was working on the division level and was stationed in Regional Command North in Mazar-i-Sharif, Afghanistan. He worked there at the request of the German military who worked with Ross and wanted an American on the NATO division-level staff. The division's area of operations spanned the northern portion of Afghanistan from the border with Pakistan, India, and China in the east in the province of Badakhshan to the border with Turkmenistan in the provinces of Faryab and Jawzjan. Within this enormous area of operations, different coalition partners had responsibility for different provinces—Germany in Kunduz and Samangan, Hungary in Baglan, Norway and Sweden in Faryab, and Mongolia in Badakhshan—and combinations of soldiers from Finland, Poland, Australia, Lithuania, Romania, Croatia, and others were intermixed around the area of operations but particularly in Mazar-i-Sharif.

Ross's unit focused mostly on supporting efforts in specific pockets of the provinces where German and Hungarian troops were primarily responsible. For Ross, this job was exciting because he was a primary planner for the Division CJ3 Operations officer, normally a position filled by a lieutenant colonel. Ross wanted to go on every mission he could, so he often represented the division as the lead operations officer while forward off base. In this role, he traveled all over the area of operations to do missions with German troops in Kunduz and Samangan, with Hungarian troops in Baglan, and Norwegian forces in Faryab.

This NATO divisional role was "an opportunity to stretch and then some," said Ross. In his new role, he was participating in working groups with multiple nations and multiple ranks, but all the allied officers were more senior than Ross. In some cases he was assigned the lead on projects working across Army, Navy, and Air Force jurisdictions from different coalition countries. Ross found the role "very rewarding." With the exception of one incident in Kunduz where he came under rocket fire from the Taliban, he had very little contact with the enemy.

When Ross returned to Fort Drum from that deployment, he took over as a company commander in the 2nd Battalion, 22nd Infantry Regiment known as "Deeds Not Words." After a year, he received a second assignment as commander of the HHC.

In January 2013 Ross deployed to Afghanistan once again. His unit was sent to Regional Command East in Afghanistan and was based in Ghazni Province. His battalion was responsible for the southern portion of the province, which included the border with Zabul leading to Kandahar in the districts of Gelan, Mukor, and Ab Bard.

Ross was the base commander at FOB Warrior, and 1st Sgt. Casey Vanwormer was the FOB "mayor." He was also the commander of the Route Clearance Package, which consisted of an infantry platoon, an engineer platoon, an explosive ordnance disposal section, a dog handler, and an air operations officer managing the air assets above.

In 2013 the United States was closing down its large network of FOBs and COPs. As a result, the distances between the remaining FOBs and COPs became very large. Indeed, they were now so large that it required a company-level mission to do IED removal between one U.S. base and the next. These missions normally took several days. To make his mission even more challenging, Ross and First Sergeant Vanwormer were in charge of the task force planning and overseeing the shutdown of their own FOB. This gave Ross a difficult task when he had to juggle his duties as base defense commander with ensuring secure ground routes to the next one, which required him to be gone for two days or more. To do this, he put in place procedures, tactics, and techniques to defend the FOB as well as many FOB-wide rehearsals that were conducted.

FOB Warrior was previously a hostile area of operations, and Ross was fortunate not to have much Taliban action to deal with. Reflecting on it years later, he recalled that "the Taliban were very calculating and spent a lot of time planning and scheming. There was always a lot of political discussions that went into their decision making. Fewer soldiers meant fewer opportunities to attack and less need if you are buying time until the enemy leaves."

In his route-clearance work, Ross would stop at every IED hotspot. He would get out of the vehicles with the infantry platoon and walk the farthest out while the engineers cleared the culverts. Ross recalls, "I would walk where we would be most vulnerable—between our vehicles and the population–where we usually got attacked from and where they [i.e., the Taliban] laced the ground with dismounted IEDs because they knew our TTPs [tactics, techniques, and procedures]. We did this because if we didn't, we would have been even more vulnerable." Ross walked slowly and carefully through areas laced with IEDs. The harsh experience of his first deployment in Iraq had taught him to be very cautious.

Ross was especially successful in his route-clearance operations. The big difference between 2007 and 2013 was that the Army had learned so much about how to combat IEDs and had much better technology. As Ross recalled, when "you look back at how many IEDs we found, it's amazing. I don't think we had any detonate on us at all. We found every single IED they laid."

On May 21, 2013, only three weeks after leaving Afghanistan, Ross returned to West Point as a TAC officer and instructor. This was a dream assignment—Ross loved West Point—and also marked the first time he was stationed somewhere for more than two years without having to deploy. In his first year back at the USMA, he studied for a master's degree through the Eisenhower Leadership Forum and also wrote a book, *Orizaba*, about his experience rescuing his best friend, Jared Sibbitt, from the tallest mountain in Mexico. Ross spent his second and third years as a TAC officer and instructor. For him, there were so many great things about being back at West Point. Everywhere he walked brought back memories of his time as a cadet. Looking at the corner of a building from a particular perspective invoked long-forgotten memories of a different day and time, happy memories of youthful cadet comradeship and fun. Medically Ross was still having some memory problems, but being able to retrace old steps helped restore it.

Professionally for Ross, the best part of being at West Point was the opportunity to teach cadets. One day in the summer of 2015, he was on his way to the Simon Center for the Professional Military Ethic and leadership carrying his résumé. He was planning to ask if he could teach PL 300. He was walking across central area when he was stopped by the head of the Simon Center, Col. Scott Halstead. He invited Ross to his office in Nininger Hall and asked if Ross would be interested in teaching MX400. This was a real honor. MX400 is a very important course and is "the Superintendent's capstone course that requires cadets to reflect upon, integrate, and synthesize their experiences in the West Point Leader Development System as they complete their development from cadet officer."[1]

Until this point, Ross did not know what MX400 was, but once he understood its significance, he accepted Colonel Halstead's offer with alacrity. For the next four semesters Ross taught MX400 and loved every moment of it.

As a combat veteran of Iraq and Afghanistan, Ross found that he had a real edge as a classroom teacher. He found it easy to get and hold cadets' attention. His cadets really appreciated learning about what he had experienced in Iraq and Afghanistan, what lessons he had learned, how he had experienced combat, and what he felt like when under enemy fire. Drawing on his experience, Ross got his cadets thinking about combat from a variety of different perspectives. For him and them it was a deeply rewarding experience.

Spending three years at West Point had another benefit for Ross, as he put it, the opportunity "to serve with some of the best leaders I have ever served with." One officer who stood out was Col. Stephen Merkel, the brigade TAC officer, "by far the most transformational and influential leader I have ever served with. The guy was amazing. He could talk to people in a way that would make them want to do anything for him to make him proud of them." Ross added that "even when he was chewing you out for doing something wrong, you did not walk away with your head down. You walked away determined to do it right next time and make him proud of you. I could not replicate that, but I sure as heck wanted to try."

Ross left West Point in the summer of 2016 bound for the Command and General Staff College at Fort Leavenworth, Kansas. His experience to date had been as a junior officer at the operational level. Taking the courses at the Command and General Staff College "opened his eyes." There he learned "how our piece of the puzzle fits into the bigger picture." Ross recalled that "there was so much that I didn't know about the Army. I had no appreciation for the greater Army—the joint aspects of military operations, how the upper echelons above brigade worked, the overall operations process, the development of strategy for warfare." He also loved learning the history: seeing how U.S. strategy evolved over time.

After graduating from the Command and General Staff College in 2017, Ross joined the "Iron Rakkasans," the 3rd Battalion, 187th Infantry, 101st Airborne Division, and served for two years before returning to Fort Benning, where he served as part of the Airborne Ranger Training Brigade and now the 198th Basic Training Brigade.

27

BOBBY SICKLER

In April 2009, Bobby Sickler returned to Fort Rucker for the Aviation Captain's Career course along with a number of his buddies from 4-6 Cav. After graduating from the course, his buddies rejoined the field army, but Bobby remained at Fort Rucker to train on the Apache.

The Apache is the most technologically complex helicopter in the world and is difficult to learn even for experienced pilots such as Bobby. It took five months to learn to fly it and another three to four months with an exceptional instructor pilot, Steve Crandall, to learn to fight with it. Crandall is the best combat pilot Bobby has ever known and taught him well. To be combat ready and confident in the aircraft as pilot in command and air mission commander took another three to four months.

On May 17, 2010, after completing his training on the Apache, Bobby joined the 82nd Air Combat Brigade at Fort Bragg, North Carolina. His mentor and former battalion commander, Col. T. J. Jamison, was the incoming brigade commander. At first, Bobby served as a staff officer for an Apache attack battalion before taking command of an Apache company in July 2011.

In September 2011 the entire brigade deployed to Afghanistan, where they were based at FOB Salerno, one of the largest U.S. military bases in southeastern Afghanistan. Because of his combat experience in Iraq, Bobby was more confident than on his first deployment. The brigade was an enlarged one with nine battalions. Bobby's battalion was a task force with one company of Apaches (which he commanded), one company of Black Hawks, one company of Chinooks, one platoon of medevac Black Hawks, one platoon of Kiowas, and the support units necessary to keep all of these aircraft flying. Within Bobby's company, only four of his sixteen pilots had flown in combat before. So, they had to rely on battalion staff senior pilots to bring greater experience to the combat crews.

The missions in Afghanistan were very similar to those Bobby had flown in Iraq but were over a much larger geographic area, and the flying in Afghanistan's high mountain ranges was more challenging and more dangerous. Bobby also flew air assault missions every night in Afghanistan, whereas he had flown only a handful of air assaults in Iraq.

Bobby's engagements with the Taliban tended to be larger and longer than those with insurgents in Iraq. The typical engagement in Iraq was very short, a five-to-ten-minute firefight against a small number of insurgents (his Distinguished Flying Cross engagement was a notable exception). In contrast, Bobby's engagements in Afghanistan against the Taliban were thirty-to-sixty-minute firefights against much larger enemy forces.

After his return to the United States in 2012, Bobby left Alpha Company to command an Apache maintenance company within the 82nd Air Combat Brigade. Just a couple of months into his command, Maj. Gen. Kevin Mangum, head of Army Aviation, appointed Bobby his personal aide. When Major General Mangum transferred out and was replaced by Maj. Gen. Michael Lundy, Bobby served as his aide for three months.

Bobby admired and respected both generals and learned a lot from them in terms of substance and style. Substantively, Bobby learned that defining a problem appropriately was most important and that limited predetermined ways of thinking can trap organizations into making bad decisions. He saw this when both generals rejected the limited options presented to them and instead chose different options entirely. Bobby also saw how commanders had a tendency to blame poor performance on their predecessors and how wrong that was. As Major General Mangum used to tell Bobby, "Never badmouth the last guy; he led a completely different unit." Finally, Bobby learned how to enhance the way he dealt with subordinates. Both generals had a way of interacting with Bobby and the rest of their staff that made them feel important and useful, and Bobby copied the generals leadership style.

Bobby's next career step was the U.S. Marine's Corps Command and General Staff College in Quantico, Virginia. There, he studied military history, security studies, leadership, and joint doctrine with electives in cyberwar and systems theory. The staff college opened a whole new world of ideas for Bobby, as he recalled:

> The heavy focus on military innovation and organizational change provided concepts and language to explain thoughts that I had started to form in Iraq and Afghanistan. The fluidity and complexity of combat never seemed to quite fit with rigid ideas about doctrine and fixed modes of operating to me. At Quantico, I was introduced to methods of thinking about change and an organization's relationship to that change that made a lot of sense to me. For the first time, I was able to see the world for the complex system that it is in ways that more closely conformed to my personal experiences.

Bobby was a distinguished graduate at the Command and General Staff College, earning a master's degree in military studies. After Quantico, Bobby moved to Hawaii, where he was XO for a Black Hawk battalion for a year and then brigade S3 for a year and finally brigade XO.

In the summer of 2018, the army told Bobby he should get a PhD but did not specify the field. Bobby chose to examine the relationship between technology and military innovation for himself and is taking a different approach than other scholars. Previous work in this field has been done by political scientists and security studies professionals. Bobby is approaching the task from a unique humanities-based perspective. There are fewer than a dozen officers in his fellowship who are addressing questions related to technology and change.

Today, in 2021, Bobby is an Advanced Strategic Planning and Policy Program fellow at the School of Advanced Military Studies studying at Arizona State University. There he is working on a PhD at the School for the Future of Innovation in Society. His dissertation will examine how best the U.S. Army can harness technology to achieve its organizational goals. This is important for the Army, which has made a significant organizational investment in trying to sort out how it can best adapt its organization as technology advances. Once he completes his doctorate, Bobby will head to U.S. Army Europe–Africa to work in its long-range planning section.

Early in 2020 as we reflected on his military career to date, I asked Bobby whether the prolonged stress and the death and destruction he had witnessed had taken a personal toll. He said it had but only during his long stay at the Walter Reed Army Medical Center. But even there in his darkest hour, he did not suffer PTSD, anxiety attacks, or nightmares. He felt guilty and useless lying in a hospital bed while other army aviation officers were in combat in Iraq and Afghanistan doing important things. Bobby was singularly focused on getting back in the fight. Once he was back in combat during his second deployment, the effect was genuinely therapeutic. He felt that he was himself again.

Like Tony Fuscellaro, an important reason Bobby did not suffer from nightmares or PTSD was because his actions as a combat pilot reflected his moral values. As Bobby told me, "I never had a bad shot. Every man I killed would have done far worse to me or my friends given half a chance." Bobby's moral fervor—intensified after witnessing a woman murdered in front of her children on his first flight in Iraq—gave him extra determination to go after insurgents whom he correctly saw as evil.

Finally, Bobby felt a sacred obligation to protect American soldiers on the ground. When he saw on U.S. television that Americans were questioning why the United States was in Iraq or Afghanistan, he thought that these debates were irrelevant. He was there in a helicopter to protect U.S. soldiers, and that was his moral duty from which he never shirked. As he took the fight to the enemy.

28

STEPHEN TANGEN

On May 4, 2011, Stephen Tangen and his soldiers from No Slack's scout platoon returned to Fort Campbell. After a year of fierce fighting, Stephen was finally out of harm's way. In July he married his fiancée, Melissa, and they enjoyed a two-week honeymoon in Hawaii after a year apart.

Before that, Lt. Col. Joel "JB" Vowell promoted Stephen to S1, battalion adjutant, and his personal executive assistant. The position brought a number of responsibilities, including battalion human resources, personnel, finance, evaluations, and the equal opportunity program. For a young officer, it was a position of great responsibility and trust.

Lieutenant Colonel Vowell wanted to appoint an outstanding adjutant because he was preparing to rotate out to become Division G3. He wanted to set his successor, Lt. Col. Brian Sullivan, up for success by ensuring he had an adjutant he could count on and trust. In appointing Stephen, Vowell paid him a great compliment. Vowell also gave Stephen an opportunity for professional growth, as the battalion adjutant's job is often the first time a junior officer gets to serve as aide-de-camp to a senior officer.

Lieutenant Colonel Sullivan was an exceptional officer appointed to battalion command in an elite division after only fourteen years in the Army. This is generally unheard of. Stephen quipped that Sullivan "looked like Xerxes from the movie *300*: six foot two with a shaved head and olive skin. Sullivan was intellectually brilliant, soft spoken, and decisive, a master delegator and leader."

Stephen found that he learned more and grew more as the battalion S1 than he had as a rifle platoon or scout platoon leader in Afghanistan. For these combat positions, he had been so well trained at West Point and Ranger School and in the yearlong train-up for his first deployment that he always knew what to do, when to do it, and why. As battalion adjutant, there was so much to learn about how the higher levels of the army worked as well as how best he could help senior officers define and solve problems. It was a more intellectually demanding role than his combat commands.

In May 2012 Stephen left No Slack for his Maneuver Captain's Careers course at Fort Benning, the next step in his professional development. He graduated in November 2012 and returned to Fort Campbell to take command of the pathfinder company 401 Aviation, 159th Combat Aviation Brigade. Before taking command, Stephen served for six months as an assistant battalion S3.

Commanding the pathfinder company was an important and challenging assignment. Pathfinder companies have an illustrious history dating back to World War II. More recently, they have served as the sole infantry company in an aviation brigade in which their missions have included securing and operating sites in hostile territory for Army helicopters and air assault troops as well as recovering downed aircraft and rescuing air crews.

The brigade was scheduled to deploy to Afghanistan in December 2013, so by the time Stephen took command, he had only four months to train up his company. What he did was condense a twelve-month training cycle into ninety days. It was intense. Every weekday, Stephen led his company through fouteen-to-eighteen-hour days of training, with usually only a few hours a night for sleep.

In designing and executing his training program, Stephen drew upon Lieutenant Colonel Vowell's vision and methods. Stephen's battalion commander, Lt. Col. Clair Gill, who observed the training closely, complimented Stephen, telling him that "this is small unit training at its finest." At his own request, Stephen took his pathfinder company to JROTC at Fort Polk, Louisiana. There he received numerous accolades for his soldiers' outstanding performance, particularly during their live-fire exercises.

On Christmas Eve 2013, Stephen deployed to Afghanistan with his pathfinder company. He had two platoons based at FOB Shank with the battalion headquarters and one platoon at the Jalalabad airfield in Nangarhar Province, just south of Kunar Province, along the Pakistani border. Stephen had to travel between the two locations, usually spending three weeks at FOB Shank and one week at Jalalabad. The company's primary mission was to recover downed aircraft and personnel.

Not long after Stephen arrived, he was briefed by Maj. Gen. Stephen Townsend, commander of Combined Joint Task Force 10 and Regional Command East, on how the commander planned to use Stephen's pathfinder company if and when he needed them. Major General Townsend also gave Stephen clear directions on what to prepare for. The three most important were to think through how best to rescue downed pilots without getting surrounded by insurgents, how to help any COP in danger of being overrun as a QRF, and rescuing missing U.S. or coalition forces.

In fact, during their deployment Stephen's company recovered some downed drones but no manned aircraft. In 2015 as the U.S. military withdrew from Afghanistan, the pathfinder company's main task was to help prepare to shut down COPs and FOBs: secure them, get the last troops out, and then leave themselves. Fortunately, there was no serious combat action during this deployment. None of Stephen's soldiers were killed or wounded. Stephen recalled that "we had a unique mission, which we executed exactly the way we were asked to, and everyone came home." It had been a learning experience that had presented its own distinct challenges, each of which Stephen had successfully met.

Stephen and his soldiers returned home in September 2014 to a difficult situation. The brigade received orders to shut down. For the next fourth months, Stephen's principal task was to get soldiers reassigned and get equipment turned in.

Stephen's next assignment was to command Charlie Company 2-11 Infantry at Fort Benning. This was a training assignment in which he was responsible for leading the officers and NCOs who instructed the IBOLC.

At this point in his career, the Army wanted Stephen to get a civilian master's degree. The Mendoza Business School at Notre Dame University offered him a fellowship, which covered the gap between what the Army paid and what the full costs would be. Leaving the field army was a culture shock, but Stephen found Notre Dame's military-friendly culture very welcoming. Unlike most civilian universities, it had a special relationship with the military dating back to World War II. Because of the wartime draft at the time and the enormous expansion of all branches of the armed services, Notre Dame was struggling to find students. Its very existence was in doubt until it leased the campus to the U.S. Navy for training. This experience wove threads of military culture into the fabric of the university. That manifests itself today not only in the annual Navy–Notre Dame football game but also in the number of retired military officers who teach at the university.

Another reason Stephen felt at home was the Mendoza Business School's strong emphasis on ethics, morals, and character, "a civilian version of West Point," as Stephen later described it. He found that the Mendoza school's goal of educating "virtuous individuals to lead human enterprises" was not mere words. Every single course he took had an ethics section. Stephen graduated in May 2018 with his West Point values strongly reinforced.

The Mendoza school also expanded Stephen's worldview and helped him understand the overlap of the structures, functions, and culture of business, government, and the military. Stephen graduated imbued with the belief that young military officers needed to think more creatively if they were to help the Army successfully address new challenges immerging in the world today and in the future.

Stephen's next challenge and the next step in his professional development was at the Command and General Staff College at Fort Leavenworth, Kansas. His experience to date had been as a junior officer at
the operational level. Taking the courses at the Command and General
Staff College "opened his eyes." Stephen had only a limited appreciation
for the greater Army: the joint aspects of military operations, how the
upper echelons above brigade worked, the overall operations process,
and the development of strategy for warfare. He also loved learning the
history, seeing how U.S. strategy evolved over time.

On a hot August day in Kansas City, I asked Stephen whether the
prolonged combat stress and the death and destruction he had witnessed
had taken a personal toll. He said it had, but he had never had nightmares, much less PTSD. Our discussion revealed that the main reason he
did not suffer from nightmares or PTSD was because his actions as a soldier reflected his own moral values, reinforced by the Army's values and
the values of the United States. He always believed that he was doing the
right thing. Morally, everything Stephen did in combat was in defense of
himself and his soldiers and was consistent with the laws of war.

CONCLUSION

At the beginning of this book, I quoted the nineteenth-century French military thinker Ardant du Picq to the effect that exceptional selfless courage on the battlefield "is the result of moral culture, and it is infinitely rare." Only elite souls possess it.

Our five elite souls—Nick Eslinger, Tony Fuscellaro, Ross Pixler, Bobby Sickler, and Stephen Tangen—never set out to be heroes or win glory. Rather, they wanted to be good, competent professionals with a strong moral compass. They did this and more, receiving numerous awards in recognition of their remarkable acts of selfless valor in Iraq and Afghanistan. They were highly effective combat leaders who demonstrated an astonishing disregard for their own lives. What mattered to them was not just accomplishing their mission but also saving the lives of their soldiers on the ground and those they protected from the air. They did so not only because it was their duty but also because they believed in the combat vision set out by their COs and because of their own deeply ingrained values.

Author David Brooks describes in his book *The Road to Character* how people such as our five elite souls act and think: "They perform acts of sacrificial service with the same modest everyday spirit they would display as if they were just getting the groceries. They are not thinking about what impressive work they are doing. They are not thinking about themselves at all."[1]

Were these elite souls' acts of selfless battlefield courage rooted in a moral culture of selflessness? Absolutely. All five demonstrated a sustained selflessness that drove their actions. In combat in Iraq and Afghanistan, they were courageous officers deeply committed to their mission, but none of them ever knowingly killed innocents.

How did our five young officers become elite souls? What made these men act so courageously on the battlefield?

Clues to answering this question are not to be found in geography, family wealth, education, or a family tradition of military service. The five elite souls came from different parts of the country: Arizona, California, Illinois, Pennsylvania, and West Virginia. Only one came from an affluent upper-middle-class home. The rest came from blue-collar, middle-class homes. Only one went to an elite private school, and it wasn't the one from the affluent home. One went to an excellent public school, and the other three went to public schools of uneven quality that did not prepare them well for West Point's high academic standards. Only one had a family tradition of citizen-soldiers. None had family who graduated from West Point.

The clue is this: all five elite souls built an unusually strong moral foundation in and through their families that was reinforced at West Point and by their COs and NCOs in the field. Our five elite souls are not perfect human beings. They would never describe themselves as such. But they are people of honor, unquestionable integrity, self-discipline, and restraint with a respect for human dignity and life and a passionate and commitment to selfless service to the nation. Unlike some military heroes, past or present, they are humble; they never boast about their achievements. They are not demigods, but they always try to do what's right.

What best helps explain their successful character building? David Brooks suggests that "example is the best teacher. Moral improvement occurs most reliably when the heart is warmed, when we come into contact with people we admire and love and we consciously and unconsciously bend our lives to mimic theirs."[2]

Each of our five elite souls grew up in a remarkable moral eco-system: devoted parents who not only taught great values but also lived them. They were models of resilience, steadfastness, and selfless devotion to family. They set high examples of what a selfless moral person looks and acts like. In each case, the parents' message was the same: always do the harder right, never lose an opportunity to serve others, and respect other people and the rule of law. Good values—duty, integrity, honor, and selflessness—help breed good character.

Their parents' lives gave each of our elite souls strong examples of resilience, steadfastness, and devotion to others. Ross's father, Reid Pixler, prosecuted Mexican drug cartels despite death threats and volunteered for the State Department's Rule of Law program in Mosul, Iraq, where he helped build a new Iraqi criminal justice system. Gina and Anthony Fuscellaro showed their son Tony the importance of serving others in times of need when the Fairless steel mill shut down and the town almost collapsed. Nick's parents both overcame difficult early lives to become models of steadfastness and selfless devotion to family. They always counseled him to do the harder right, the selfless not the selfish thing. Bobby Sickler's father served twenty-six years as a Marine Corps officer and tutored his son on moral virtue. Stephen Tangen's parents gave him daily lessons in service to others and personal resilience despite punishing hours as a surgical nurse or being betrayed by a business partner, which compromised the family economically. For four of them, character was also shaped by active engagement in a close network of Christian churches, where their parents also played active roles.

Youth organizations led by selfless coaches and mentors were also important. For Nick and Stephen, sports coaches taught them to perform with honor and integrity; cheating and foul play were unacceptable. For Bobby and Ross, outstanding Japanese karate coaches helped them grasp and practice the sport's three standards of honor: obligation, justice, and courage. Ross and Stephen were Eagle Scouts, and Ross was a JROTC commander. And best friends helped reinforce moral values.

By age eighteen, all five elite souls were not only accomplished young leaders but also young men with an exceptionally strong moral foundation. All five had strong moral reasons for wanting to attend

West Point and embark on a military career. All five had given careful thought to the USMA's honor code and honor system and tried to internalize it. Ross, for example, thought of the USMA in almost religious terms, seeing it as the way for him to answer his calling. In his West Point application, Stephen wrote that his honor code and West Point's were identical. Bobby respected West Point's clear-cut honor code. Nick admired the moral character of the West Point cadets who visited his high school. Tony highlighted the compatibility of his moral values with those of West Point.

West Point's core mission is to produce servant leaders of character. The USMA took these five outstanding young men, broke down their individual egos, and rebuilt them. But that transformation was a slow, grueling process, and all of the elite souls struggled at one point or another. Most struggled academically, one of them almost failed Beast Barracks, and another violated the honor code. West Point made them think less as individuals and more so of their role within the Army team. The USMA planted the idea of the preeminence of the team over the individual during Cadet Basic Training and developed this idea through its Leader Development Program. Through this transformation, West Point built upon their strong moral foundation, helped them develop their character further, and brought out in them the core values that unite the military team: duty, honor, integrity, loyalty, respect, selflessness, and personal courage. Put simply, West Point lifted our five young officers, who already had an unusually strong moral foundation, to even greater heights.

The central argument of this book has been that each of the five elite souls—Nick Eslinger, Tony Fuscellaro, Ross Pixler, Bobby Sickler, and Stephen Tangen—embody a marriage of outstanding ability with exceptional moral character. Each of them entered West Point as young men of great promise with impressive records of intellectual leadership and physical achievements. Each brought a moral grounding secured by exemplary parents, coaches, and friends. This is not to suggest that these five officers are demigods or superheroes. They are human, very human. Later, two had moments of serious self-doubt at Ranger School. In Iraq, one was investigated for shooting an armed insurgent who was running away. Whatever its flaws, West Point's rigorous educational, military training and development model built on their strong early foundation to produce young military leaders who personified the classical virtues of moral and physical courage, selflessness, a deep commitment to duty, personal and professional honor, and selfless service to the nation. They are among the very best that West Point produces.

NOTES

INTRODUCTION

1. John Ellis, quoted in Paul Fussell, *Wartime* (New York: Oxford University Press, 1989), 281, 285.
2. Marc Aronson and Patty Campbell, eds., *War Is* (Somerville, MA: Candlewick, 2008), 13.
3. Lord Moran, *The Anatomy of Courage* (New York: Carroll and Graf, 2007), xii.
4. Elizabeth D. Samet, *Soldier's Heart: Reading Literature through Peace and War at West Point* (New York: Farrar, Straus and Giroux, 2008), 188.
5. Sebastian Junger, *War* (New York: Hachette, 2010), 211.
6. Charles Jean Jacques Joseph Ardant du Picq, *Battle Studies*, trans. and ed. Roger J. Spiller (Lawrence: University Press of Kansas, 2017), 40–41.
7. Samet, *Soldier's Heart*, 232.

CHAPTER 1. NICK ESLINGER: EARLY LIFE

1. From one of Nick's application essays for West Point, November 12, 2001, Nick Eslinger, Application file, USMA Archives.
2. Ibid.
3. Kevin Allen, letter of recommendation for West Point, November 15, 2001, Nick Eslinger, Application file, USMA Archives.
4. Mark Gates, letter of recommendation for West Point, November 7, 2001, Nick Eslinger, Application file, USMA Archives.
5. Jeff Jonas, letter of recommendation for West Point, October 31, 2001, Application file, USMA Archives.

6. Jeff Jonas, letter of recommendation for West Point, October 31, 2001, Application file, USMA Archives.

CHAPTER 2. TONY FUSCELLARO: EARLY LIFE

1. James T. Patterson, *Grand Expectations: The United Stars, 1945–1971* (New York: Oxford University Press, 1996), 70–74.
2. Anthony Figliola, letter of recommendation for West Point, undated, most likely September 2000, Anthony Fuscellaro Application file, USMA Archives.
3. Jeffrey Danilak, letter of recommendation for West Point, September 21, 2000, Anthony Fuscellaro Application file, USMA Archives.
4. Anthony Figliola, letter of recommendation for West Point, undated, most likely September 2000, Anthony Fuscellaro Application file, USMA Archives.
5. Ibid.
6. Application essay for West Point, October 27, 2000, Anthony Fuscellaro Application file, USMA Archives.

CHAPTER 3. ROSS PIXLER: EARLY LIFE

1. Ross Carlos Pixler, *Orizaba: The Extraordinary Story of Two Mountain Climbers' Struggle for Survival* (New York: Pylon Publishing, 2015), 45.
2. Ibid., 32.
3. Application essay for West Point, December 18, 2000, Ross Pixler Application File, USMA Archives.
4. Ibid.
5. For a full account, see Pixler, *Orizaba*.

CHAPTER 4. BOBBY SICKLER: EARLY LIFE

1. Susan Baird, interview notes, Bobby Sickler, July 18, 2000, Bobby Sickler Application File, USMA Archives.
2. Marsha Keller, letter of recommendation for West Point, August 25, 2000, Bobby Sickler Application File, USMA Archives.
3. Susan Baird, interview notes, Bobby Sickler, July 18, 2000, Bobby Sickler Application File, USMA Archives.
4. Ibid.
5. Application essay for West Point, August 25, 2000, Bobby Sickler Application File, USMA Archives.

CHAPTER 5. STEPHEN TANGEN: EARLY LIFE

1. Laura Sova Hoglund, letter of recommendation for West Point, undated but most likely September 2003, Stephen Tangen Application File USMA Archives.
2. Eugene Pecker, letter of recommendation for West Point, September 3, 2003, Stephen Tangen Application File, USMA Archives.
3. Application essay for West Point, September 1, 2003, Stephen Tangen Application File, USMA Archives.

CHAPTER 6. WEST POINT

1. David Brooks, *The Second Mountain: The Quest for a Moral Life* (New York: Random House, 2019), 194.
2. "USMA Medal of Honor Recipients," West Point Association of Graduates, n.d., www.westpointaog.org/file/WestPointGraduateMedalofHonor Recipients.pdf.
3. Email to the author, April 18, 2019.
4. Ibid.
5. David Lipsky. *Absolutely American* (New York: Vintage Books, 2004), 34.
6. Email to the author, November 10, 2020.
7. Lance Betros, *Carved from Granite: West Point since 1902* (College Station: Texas A&M University Press, 2012), 263.
8. Ibid., 289.
9. Ibid., 290.
10. Ibid.
11. Ibid., 291.
12. Ibid., 291–94.
13. Ibid., 293–94.
14. Ibid.
15. Ibid.
16. Ibid.
17. Ibid., 295.
18. Ibid.
19. Ibid., 296.
20. Ibid., 73.
21. Ibid., 88.
22. Ibid., 96.
23. Ibid.

24. Ibid.
25. J. Pepper Bryars, *American Warfighter: Brotherhood, Survival and Uncommon Valor in Iraq, 2003–2011* (Self-published, 2016), 290.
26. Betros, *Carved from Granite*, 89, 100.
27. David Brooks, *The Road to Character* (New York: Random House, 2016), 13.
28. Ed Ruggero, *Duty First: A Year in the Life of West Point and the Making of American Leaders* (New York: Harper Perennial, 2002), 9.
29. Craig Mullaney, *The Unforgiving Minute: A Soldier's Education* (New York: Penguin 2009), 48.
30. Betros, *Carved from Granite*, 239–40.
31. Ibid., 262.
32. Ibid., 257.
33. Brig. Gen. (Ret.) Jim Golden, email to the author, September 23, 2019.
34. Betros, *Carved from Granite*, 257–58.
35. Ruggero, *Duty First*, 29–34.
36. Ibid., 200–220.
37. Ibid., 161.
38. Ibid., 34.
39. Betros, *Carved from Granite*, 210.
40. Ibid., 209.
41. Ibid., 224–38.
42. Ibid., 223–24.
43. Ibid., 233–34, 236–37.
44. Ibid., 231.
45. Ibid., 232.
46. Ibid., 233.
47. Ibid., 235.
48. Bruce Keith, "The Transformation of West Point as a Liberal Arts College," *Liberal Education* 96, no. 2 (2010): 1.
49. Betros, *Carved from Granite*, 111–61.
50. Ibid., 136–39.
51. Ibid., 141.
52. Ibid., 147.
53. Ibid., 146–47.
54. Ibid., 147.
55. Ibid., 150–52.
56. Ibid., 152.
57. Ibid.
58. Ibid., 153.
59. Ibid.

60. Ibid., 155.
61. Ibid., 155–56.
62. Email to the author, October 28, 2020.
63. Email to the author, November 2, 2020.
64. Ibid.
65. Email to the author, October 28, 2020.
66. Ibid.
67. Ibid.
68. Email to the author, October 27, 2020.
69. Email to the author, February 5, 2021.

CHAPTER 9. ROSS PIXLER: USMA

1. U.S. Army, "Air Assault School," https://m.goarmy.com/soldier-life/being-a-soldier/ongoing-training/specialized-schools/air-assault.m.html.
2. Ibid.
3. Ibid.

CHAPTER 10. BOBBY SICKLER: USMA

1. U.S. Army, "Air Assault School," https://m.goarmy.com/soldier-life/being-a-soldier/ongoing-training/specialized-schools/air-assault.m.html.
2. Ibid.

CHAPTER 12. RANGER SCHOOL

1. John Spencer, "Ranger School Is Not a Leadership School," Real Clear Defense, December 5, 2016, https://www.realcleardefense.com/articles/2016/12/06/ranger_school_is_not_a_leadership_school_110444.html.
2. Stew Smith, "Preparing for Army Ranger School," Military.com, n.d., https://www.military.com/military-fitness/army-special-operations/army-ranger-school-prep.
3. "Student Information," Fort Benning, n.d., https://www.benning.army.mil/infantry/artb/Student-Information/.

CHAPTER 16. FORT RUCKER

1. "Initial Entry Rotary Wing Training," U.S. Army Aviation Center of Excellence and Fort Rucker, n.d., https://home.army.mil/rucker/index.php/about/usaace/student-information/ierw.

CHAPTER 17. TONY FUSCELLARO: PREPARATION

1. Amber Smith, *Danger Close: My Epic Journey as a Combat Helicopter Pilot in Iraq and Afghanistan* (New York: Atria, 2016), 35.

CHAPTER 18. BOBBY SICKLER: PREPARATION

1. Michael L. Wesolek, "Analysis of the Effectiveness of Army Helicopter Training," *Journal of Aviation/Aerospace Education and Research* 18, no. 2 (2009): 69.

CHAPTER 19. NICK ESLINGER: IRAQ, 2008–2009

1. Michal Harari, "Uncertain Future of the Sons of Iraq," Institute for the Study of War, August 3, 2010, https://www.understandingwar.org/sites/default/files /Backgrounder_SonsofIraq_0.pdf. Harari is quoting Farook Ahmed, "Sons of Iraq and the Awakening Forces," Institute for the Study of War, February 21, 2008, https://www.understandingwar.org/sites/default/files/reports /Backgrounder%2023%20Sons%20of%20Iraq%20and%20Awakening %20Forces.pdf.
2. For an excellent analysis of the US Army's use of T-walls in the urban environments of Iraq, see John Spencer, "The Most Effective Weapon on the Modern Battlefield Is Concrete," Real Clear Defense, November 14, 2016, https://www.realcleardefense.com/articles/2016/11/15/the_most_effective _weapon_on_the_modern_battlefield_is_concrete_110348.html.

CHAPTER 20. TONY FUSCELLARO: AFGHANISTAN, 2009–2010

1. Kate Wiltrout, "Virginia-Bred Hero Awarded Silver Star for Brave Piloting," *Roanoke Times*, July 3, 2010, https://roanoke.com/archive/virginia-bred-hero -awarded-silver-star-for-brave-piloting/article_dc58768f-cb49-51b4-acfc -e903702d8f98.html.
2. Quoted in Karl Hawkins, "Kiowa Team Gets 'Danger Close' to Save Lives," U.S. Army, May 6, 2011, https://www.army.mil/article/56103/kiowa_team _gets_danger_close_to_save_lives.
3. The Dushkas were heavily used by the Soviet military in Afghanistan between 1979 and 1989. It is possible that the mujahideen, a forerunner of the Taliban, captured Dushkas from the Soviets or found ones abandoned on former Soviet bases. Some squadron Kiowas had been hit by 12.7-millimeter rounds from the Dushkas.

4. Kyle Mizokami, in his article "The Army's Recoilless Rifle Is Getting a Huge Upgrade: Laser-Guided Warheads," *Popular Mechanics*, October 10, 2018, explained what the recoilless rifle is: "The propellant gasses are directed backward, counteracting the weapon's recoil, making it 'recoilless.' The weapon, which is basically a hollow tube with a grip stock and trigger, has spiral rifling on the barrel to impart spin stabilization on the round as it exits the barrel. Hence the name recoilless rifle."

CHAPTER 21. ROSS PIXLER: IRAQ, 2007–2008

1. Philip Caputo, *A Rumor of War* (New York: Ballantine Books 1977), xvii.

CHAPTER 22. BOBBY SICKLER: IRAQ, 2007–2008

1. Michael Yon, "Guitar Heroes," Michael Yon Online Magazine, March 10, 2008, https://www.michaelyon-online.com/guitar-heroes.htm.

CHAPTER 23. STEPHEN TANGEN: AFGHANISTAN, 2010

1. Philip Caputo, *A Rumor of War* (New York: Ballantine Books, 1978), xvii.
2. This chapter is based on several extended interviews with Maj. Stephen Tangen and on his paper "Operation Strong Eagle," Maneuver Captains Career Course 4-12, June 27–29, 2010. The chapter also draws on John McGrath, "Operation Strong Eagle: Combat Action in the Ghaki Valley," in *Vanguard of Valor*, ed. Donald P. Wright, 91–131 (Fort Leavenworth, KS: U.S. Army Combined Arms Center, Combined Studies Institute Press, 2011).
3. Tangen, "Operation Strong Eagle," 1; McGrath, "Operation Strong Eagle," 94.
4. McGrath, "Operation String Eagle," 92.
5. Ibid.
6. Ibid., 93–94; Tangen, "Operation Strong Eagle," 2.
7. Tangen, "Operation Strong Eagle," 1.
8. Bill Roggio, "The Most Wanted Taliban Commanders in Afghanistan," DD's Long War Journal, April 11, 2011, https://www.longwarjournal.org/archives/2011/04/most_hunted_the_most.php.
9. Bill Roggio, "Afghan Military Claims Dual-Hatted Taliban and al-Qaeda Leader Killed in ISAF Airstrike," FDD's Long War Journal, August 22, 2013, https://www.longwarjournal.org/archives/2013/08/afghan_military_clai.php.
10. Ibid.
11. Interview in *No Greater Love*, directed by Jason Roberts (Atlas, 2015).

12. McGrath, "Operation Strong Eagle," 100–101.
13. Tangen, "Operation Strong Eagle," 4–5.
14. McGrath, "Operation Strong Eagle," 104.
15. Ibid.
16. Ibid.
17. Ibid.
18. Ibid.
19. The MATV is the lighter all-terrain version of the MRAP. The MaxxPro is the heavier version. Both were designed with a V-shaped internal crew compartment built on a strong International 700 chassis. The idea was that this would deflect the explosive from a mine or IED away from the soldiers inside. Both vehicle types were fresh from their U.S. factory thanks to the tireless efforts of Defense Secretary Robert Gates.
20. Tangen, "Operation Strong Eagle," 5.
21. Ibid., 6.
22. Ibid., 9.
23. Ibid.
24. Ibid., 8.
25. Deployed in Afghanistan since 2003, the Buffalo was a large heavily armored mine detection and removal vehicle fourteen feet tall by eight feet wide. With a crew of six, the Buffalo could dispose of IEDs and mines with a thirty-foot robotic iron claw that could be operated from the safety of the armored cabin.
26. Tangen, "Operation Strong Eagle," 10.
27. Ibid., 11.
28. Ibid.
29. Ibid., 12.
30. Ibid.
31. Ibid., 12–13.
32. Ibid., 13.
33. Ibid., 13–14.
34. Ibid., 14.
35. Ibid.
36. Ibid.
37. Ibid., 15.
38. Ibid.
39. Ibid., 16.
40. Ibid.
41. Ibid., 17.

42. Ibid.
43. Ibid.
44. Ibid., 18.
45. Ibid., 18–19.
46. Ibid., 19.
47. Ibid.
48. Ibid., 20.
49. Ibid., 21.
50. Ibid.
51. Ibid., 21–22.
52. Ibid., 22.
53. Ibid., 24.
54. Ibid.
55. Ibid.

CHAPTER 24. NICK ESLINGER

1. Quoted in Katie Peterson, "Award Upgraded to Distinguished Service Cross," *Fort Leavenworth Lamp*, May 8, 2019, https://www.ftleavenworthlamp.com /news/top-news-stories/2019/05/09/award-upgraded-to-distinguished-service -cross/.

CHAPTER 26. ROSS PIXLER

1. "MX400 Course Details," Westpoint, n.d., https://courses.westpoint.edu /crse_details.cfm?crse_nbr=MX400&int_crse_eff_acad_yr=2018&int_crse _eff_term=5.

CONCLUSION

1. David Brooks, *The Road to Character* (New York: Random House, 2015), xvi.
2. Ibid. xv.

SELECTED BIBLIOGRAPHY

Aronson, Marc, and Patty Campbell, eds. *War Is.* Somerville, MA: Candlewick, 2009.

Betros, Lance. *Carved from Granite: West Point since 1902.* College Station: Texas A&M University Press, 2012.

Bolger, Daniel P. *Why We Lost: A General's Inside Account of the Iraq and Afghanistan Wars.* New York: Mariner, 2015.

Brooks, David. *The Road to Character.* New York: Random House, 2015.

———. *The Second Mountain: The Quest for a Moral Life.* New York: Random House, 2019.

Bryars, J. Pepper. *American Warfighter: Brotherhood, Survival, and Uncommon Valor in Iraq, 2003–2011.* Self-published, 2016.

Chivers, C. J. *The Fighters: Americans in Combat in Afghanistan and Iraq.* New York: Simon and Schuster, 2018.

Christensen, Loren W., and Dave Crossman. *On Combat: The Psychology and Physiology of Deadly Conflict in War and in Peace.* Mascoutah, IL: Warrior Science Publications, 2008.

Couch, Dick. *The Sheriff of Ramadi: Navy SEALs and the Winning of al-Anbar.* Annapolis, MD: Naval Institute Press, 2008.

Cowley, Robert, and Thomas Guinzburg. *West Point: Two Centuries of Honor and Tradition.* New York: Warner Books, 2002.

Crackel, Theodore J. *The Illustrated History of West Point.* New York: Harry N. Abrams, 1991.

Finkel, David. *The Good Soldiers.* New York: Sarah Crichton, 2009.

Grant, John, James Lynch, and Ronald Bailey. *West Point: The First Two Hundred Years.* Guilford, CT: Globe Pequot, 2002.

Gray, J. Glenn. *The Warriors: Reflections on Men in Battle.* New York: Harper and Row, 1970.

Junger, Sebastian. *War.* New York: Hachette, 2010.

Kennedy, Kelly. *They Fought for Each Other: The Triumph and Tragedy of the Hardest Hit Unit in Iraq.* New York: St. Martin's Griffin, 2010.

Lipsky, David. *Absolutely American: Four Years at West Point.* New York: Vintage, 2004.

Mansoor, Peter R. *Baghdad at Sunrise: A Brigade Commander's War in Iraq.* New Haven, CT: Yale University Press, 2008.

Meyer, Dakota, and Bing West. *Into the Fire: A Firsthand Account of the Most Extraordinary Battle in the Afghan War.* New York: Random House, 2012.

Moran, Lord (Sir Charles Watson). *The Anatomy of Courage.* New York: Carroll and Graff, 2007.

Mullaney, Craig. *The Unforgiving Minute: A Soldier's Education.* New York: Penguin, 2009.

Murphy, Bill, Jr. *In a Time of War: The Proud and Perilous Journey of West Point's Class of 2002.* New York: Henry Holt, 2008.

Patterson, James T. *Grand Expectations: The United Stars, 1945–1971.* New York: Oxford University Press, 1996.

Picq, Ardant du. *Battle Studies.* Lawrence: University Press of Kansas, 2017.

Pixler, Ross. *Orizaba: The Extraordinary Story of Two Mountain Climbers' Struggle for Survival.* New York: Pylon Publishing, 2015.

Ricks, Thomas E. *Fiasco: The American Military Fiasco in Iraq.* New York: Penguin, 2007.

Robinson, Linda. *Tell Me How This Ends: General David Petraeus and the Search for a Way Out of Iraq.* New York: PublicAffairs, 2008.

Romesha, Clinton. *Red Platoon: A True Story of American Valor.* New York: Dutton, 2018.

Ruggero, Ed. *Duty First: A Year in the Life of West Point and the Making of American Leaders.* New York: Harper Perennial, 2001.

Samet, Elizabeth D. *Soldier's Heart: Reading Literature through Peace and War at West Point.* New York: Farrar, Straus and Giroux, 2008.

Sherman, Nancy. *The Untold War: Inside the Hearts, Minds, and Souls of Our Soldiers.* New York: Norton, 2010.

Smith, Amber. *Danger Close: My Epic Journey as a Combat Helicopter Pilot in Iraq and Afghanistan.* New York: Atria, 2016.

Smith, Larry. *Medal of Honor.* New York: Norton, 2003.

Stephenson, Michael. *The Last Full Measure: How Soldiers Die in Battle.* New York: Crown, 2012.

INDEX

ABOUT THE AUTHOR

RAYMOND JAMES RAYMOND is a former British diplomat. He is an adjunct professor in the Department of Social Sciences, United States Military Academy; adjunct fellow of the Pell Center for International Relations and Public Policy, Newport, Rhode Island; and professor emeritus of government and history at the State University of New York campus at Stone Ridge. He is currently a visiting professor of government at Masaryk University in the Czech Republic.

The Naval Institute Press is the book-publishing arm of the U.S. Naval Institute, a private, nonprofit, membership society for sea service professionals and others who share an interest in naval and maritime affairs. Established in 1873 at the U.S. Naval Academy in Annapolis, Maryland, where its offices remain today, the Naval Institute has members worldwide.

Members of the Naval Institute support the education programs of the society and receive the influential monthly magazine *Proceedings* or the colorful bimonthly magazine *Naval History* and discounts on fine nautical prints and on ship and aircraft photos. They also have access to the transcripts of the Institute's Oral History Program and get discounted admission to any of the Institute-sponsored seminars offered around the country.

The Naval Institute's book-publishing program, begun in 1898 with basic guides to naval practices, has broadened its scope to include books of more general interest. Now the Naval Institute Press publishes about seventy titles each year, ranging from how-to books on boating and navigation to battle histories, biographies, ship and aircraft guides, and novels. Institute members receive significant discounts on the Press' more than eight hundred books in print.

Full-time students are eligible for special half-price membership rates. Life memberships are also available.

For more information about Naval Institute Press books that are currently available, visit www.usni.org/press/books. To learn about joining the U.S. Naval Institute, please write to:

Member Services
U.S. Naval Institute
291 Wood Road
Annapolis, MD 21402-5034
Telephone: (800) 233-8764
Fax: (410) 571-1703
Web address: www.usni.org